METHODS IN MOLECULAR BIOLOGY

Series Editor
John M. Walker
School of Life and Medical Sciences
University of Hertfordshire
Hatfield, Hertfordshire, AL10 9AB, UK

For further volumes:
http://www.springer.com/series/7651

Plant Nitric Oxide

Methods and Protocols

Edited by

Kapuganti Jagadis Gupta

National Institute of Plant Genome Research, New Delhi, India

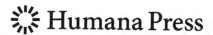 Humana Press

Editor
Kapuganti Jagadis Gupta
National Institute of Plant Genome Research
New Delhi, India

ISSN 1064-3745   ISSN 1940-6029   (electronic)
Methods in Molecular Biology
ISBN 978-1-4939-8090-1   ISBN 978-1-4939-3600-7   (eBook)
DOI 10.1007/978-1-4939-3600-7

This Humana Press imprint is published by Springer Nature
The registered company is Springer Science+Business Media LLC New York

Preface

After nearly one and half decades of study on nitric oxide (NO) research, we are now clear that NO is one of the major signal molecules that play a role in plant growth development and stress physiology. In order to understand the role of NO, it is very important to precisely quantify NO. Due to a very low half-life and high reactivity, it is very difficult to precisely quantify NO. This problem is further complicated by multiple sources of NO generation in plants. Free radicals play an important role in scavenging NO; therefore cell conditions play a role in the determination of NO.

Nearly a dozen methods are currently available for the quantification of NO. Each method has several advantages and disadvantages. This book will provide a foundation for all nitric oxide researchers to choose the required method for their research work.

For a new researcher who plans to measure NO in plant materials, it may be difficult to select a method from a number of methods. In this book, Yamasaki et al. provide a practical guide on the choosing of a technique for measuring NO from plant materials.

Chemiluminescence is one of the reliable methods in which nitric oxide reacts with ozone and creates chemiluminescence. In this method, one can detect the gaseous form of NO that is released into the atmosphere. Due to high reactivity, NO can be oxidized, and it can be detected by indirect chemiluminescence. The sum of NO released from direct and indirect chemiluminescence can reflect the total amount of NO. The detailed overview of this method is presented by Wany et al. in the second chapter. Chemiluminescence can be used to detect NO from purified enzymes, and it is also possible to detect scavenging of NO by various cell components. Mishra et al. explore these procedures.

The most widely used method for the measurement of NO is the diaminofluorescence method. The excellent advantage of DAF-based dyes is due to its easy applicability by pipette to the tissue of interest and observation of the emitted NO fluorescence using any laboratory fluorescence and confocal microscope. Site-specific NO production (localization) is important in many studies. DAF is the best dye for such studies. Sensitivity of these dyes is in Nano molar range, and no additional fluorescence is observed with several reactive oxygen and nitrogen species such as NO_2^-, NO_3^-, H_2O_2, and $ONOO^-$. Wany and Gupta present a stepwise protocol for localization of NO in roots. Agurla et al. present a method for the detection of NO in guard cells using a DAF-2DA method.

As with every method, DAF has several disadvantages too. In this regard, Ruemer et al. explain that DAF dyes do not only react with nitric oxide but also react with peroxidase enzyme and hydrogen peroxide. Since DAF has several disadvantages, an alternate solution obtained by Jain and Bhatla involves a simple, two-step synthesis, characterization, and application of MNIP-Cu {Copper derivative of [4-methoxy-2-(1H-napthol [2,3-*d*] imidazol-2-yl)phenol]} for specific and rapid binding with NO, which can be detected by epifluorescence microscopy and confocal laser scanning microscopy (CLSM). The advantage is that it can detect NO under normoxic and anoxic conditions.

There are also several other methods available. One of the next reliable methods is EPR spectroscopy. In this context, Maia and Moura chapter provide comprehensive information about EPR spectroscopy and of some spin-trapping methodologies to study NO. They also discuss "strengths and weaknesses" of iron-dithiocarbamates utilization, the NO traps, and also provide a detailed description of the method to quantify the NO formation by molyb-doenzymes. Galatro and Susana next provide a stepwise protocol for the measurement of NO from chloroplasts.

Mandon et al. provide extensive information on a laser-based method for the detection of nitric oxide. This works on a principle in which changes in light intensities or polarization occurs when the laser light is interacting with NO molecules. This laser-based photo-acoustic method is very popular. The molecules absorb the energy from the laser light, and a pressure wave (sound wave) is generated due to their thermal expansion. The amplitude of this sound wave, proportional to the amount of molecules, is detected by a sensitive microphone.

The Griveau et al. chapter describes the principle, the preparation, and the use of a home-made electrode displaying a high specificity for NO detection in plant cell suspensions. This chapter presents an exhaustive introduction regarding NO measurement in plants and the use of electrodes as an alternative method. This allows the reader to understand the problems related to the detection of NO due to its high reactivity.

The Barroso et al. article provides two complementary approaches which can be extensively useful for studying the content and distribution of S-nitrosothiols in different plant tissues and species under various conditions.

The Noelia et al. chapter describes detailed protocols to study the expression and characterization of the enzymatic activity of NOS from *O. tauri*. The authors demonstrate NOS activity using an oxyhemoglobin assay, citrulline assays, and the NADPH oxidation for in vitro analysis, fluorescent probes, and Griess assay for in vivo determination. This chapter further discusses the advantages and drawbacks of each method. Loake et al. describe a procedure for the identification of S-nitrosothiols to study the roles of protein S-nitrosylation in the immune responses of *Arabidopsis thaliana* and other organisms. This technique employs a modified version of the biotin-switch technique, which we termed the sequential cysteine blocking (SCB) technique, encompassing the sequential redox-blocking of recombinant proteins followed by LC-MS/MS analysis.

S-nitrosoglutathione reductase (GSNOR) is considered a key enzyme in the regulation of intracellular levels of S-nitrosoglutathione and protein S-nitrosylation. Kubienová presents optimized protocols to determine GSNOR enzyme activities by spectrophotometry coupled with activity staining after the native polyacrylamide gel electrophoresis. Peroxynitrite is formed in a reaction between nitrite oxide and superoxide, and the detection of this compound is crucial in various situations such as plant pathogen interactions and nitrosative stress. Bellin et al. described a detailed method to measure peroxynitrite.

This methods book contains detailed information about methods in plant nitric oxide research and is very helpful for all researchers working and intending to work on plant nitric oxide research.

Reputed scientists from 11 different counties have contributed to this book to whom I am extremely thankful. I thank Aprajita Kumari for help in the formatting of the manuscripts, and I owe my heartfelt gratitude to John Walker for his support and timely advice during the preparation of this book.

New Delhi, India   *Kapuganti Jagadis Gupta*

Contents

Contributors

SRINIVAS AGURLA • *Department of Plant Sciences, School of Life Sciences, University of Hyderabad, Hyderabad, India*

JUAN B. BARROSO • *Group of Biochemistry and Cell Signaling in Nitric Oxide, Department of Biochemistry and Molecular Biology, University of Jaén, Jaén, Spain*

FETHI BEDIOUI • *Unité de Technologies Chimiques et Biologiques pour la Santé, Chimie ParisTech, PSL Research University, Paris, France; Unité de Technologies Chimiques et Biologiques pour la Santé UMR 8258, CNRS, Paris, France; Unité de Technologies Chimiques et Biologiques pour la Santé, Université Paris Descartes, Paris, France; Unité de Technologies Chimiques et Biologiques pour la Santé (N°1022), INSERM, Paris, France*

JUAN C. BEGARA-MORALES • *Group of Biochemistry and Cell Signaling in Nitric Oxide, Department of Biochemistry and Molecular Biology, University of Jaén, Jaén, Spain*

DIANA BELLIN • *Dipartimento di Biotecnologie, Università degli Studi di Verona, Verona, Italy*

ANGÉLIQUE BESSON-BARD • *Université de Bourgogne, UMR 1347 Agroécologie, Dijon, France; ERL CNRS 6300, Dijon, France*

SATISH C. BHATLA • *Department of Botany, University of Delhi, Delhi, India*

THIERRY LE BIHAN • *Synthetic and Systems Biology, School of Biological Sciences, University of Edinburgh, Edinburgh, UK*

ALFONSO CARRERAS • *Group of Biochemistry and Cell Signaling in Nitric Oxide, Department of Biochemistry and Molecular Biology, University of Jaén, Jaén, Spain*

MOUNIRA CHAKI • *Group of Biochemistry and Cell Signaling in Nitric Oxide, Department of Biochemistry and Molecular Biology, University of Jaén, Jaén, Spain*

MICHAEL F. COHEN • *Department of Biology, Sonoma State University, Rohnert Park, CA, USA; Biological Systems Unit, Okinawa Institute of Science and Technology, Okinawa, Japan*

FRANCISCO J. CORPAS • *Group of Antioxidants, Free Radicals and Nitric Oxide in Biotechnology, Food and Agriculture, Department of Biochemistry, Cell and Molecular Biology of Plants, Estación Experimental del Zaidín, CSIC, Granada, Spain*

NATALIA CORREA-ARAGUNDE • *Instituto de Investigaciones Biológicas, Facultad de Ciencias Exactas y Naturales, Universidad Nacional de Mar del Plata, Mar del Plata, Argentina*

SIMONA M. CRISTESCU • *Department of Molecular and Laser Physics, Radboud University Nijmegen, Nijmegen, The Netherlands*

ANISHA DAVID • *Department of Botany, University of Delhi, Delhi, India*

MASSIMO DELLEDONNE • *Dipartimento di Biotecnologie, Università degli Studi di Verona, Verona, Italy*

AGNES FEKETE • *Julius-von-Sachs Institute of Biosciences, University of Wuerzburg, Wuerzburg, Germany*

NOELIA FORESI • *Instituto de Investigaciones Biológicas, Facultad de Ciencias Exactas y Naturales, Universidad Nacional de Mar del Plata, Mar del Plata, Argentina*

ANDREA GALATRO • *Physical Chemistry-Institute of Biochemistry and Molecular Medicine (IBIMOL), School of Pharmacy and Biochemistry, University of Buenos Aires-CONICET, Buenos Aires, Argentina*

GUNJA GAYATRI • *Department of Plant Sciences, School of Life Sciences, University of Hyderabad, Hyderabad, India*

SOPHIE GRIVEAU • *Unité de Technologies Chimiques et Biologiques pour la Santé, Chimie ParisTech, PSL Research University, Paris, France; Unité de Technologies Chimiques et Biologiques pour la Santé UMR 8258, CNRS, Paris, France; Unité de Technologies Chimiques et Biologiques pour la Santé, Université Paris Descartes, Paris, France; Unité de Technologies Chimiques et Biologiques pour la Santé (N°1022), INSERM, Paris, France*

ALOK KUMAR GUPTA • *National Institute of Plant Genome Research, New Delhi, India*

KAPUGANTI JAGADIS GUPTA • *National Institute of Plant Genome Research, New Delhi, India*

SHIKA GUPTA • *National Institute of Plant Genome Research, New Delhi, India*

FRANS J.M. HARREN • *Department of Molecular and Laser Physics, Radboud University Nijmegen, Nijmegen, The Netherlands*

RAFAEL A. HOMEM • *Institute of Molecular Plant Sciences, School of Biological Sciences, University of Edinburgh, Edinburgh, UK*

PRACHI JAIN • *Department of Botany, University of Delhi, Delhi, India*

RITIKA JAINTU • *National Institute of Plant Genome Research, New Delhi, India*

WERNER M. KAISER • *Julius-von-Sachs Institute of Biosciences, University of Wuerzburg, Wuerzburg, Germany*

MARKUS KRISCHKE • *Julius-von-Sachs Institute of Biosciences, University of Wuerzburg, Wuerzburg, Germany*

LUCIE KUBIENOVÁ • *Department of Biochemistry, Faculty of Science, Palacky University, Olomouc, Czech Republic*

APRAJITA KUMARI • *National Institute of Plant Genome Research, New Delhi, India*

LORENZO LAMATTINA • *Instituto de Investigaciones Biológicas, Facultad de Ciencias Exactas y Naturales, Universidad Nacional de Mar del Plata, Mar del Plata, Argentina*

MARIA LESCH • *Julius-von-Sachs Institute of Biosciences, University of Wuerzburg, Wuerzburg, Germany*

GARY J. LOAKE • *Institute of Molecular Plant Sciences, School of Biological Sciences, University of Edinburgh, Edinburgh, UK; Synthetic and Systems Biology, School of Biological Sciences, University of Edinburgh, Edinburgh, UK*

LENKA LUHOVÁ • *Department of Biochemistry, Faculty of Science, Palacky University, Olomouc, Czech Republic*

LUISA B. MAIA • *UCIBIO, REQUIMTE, Departamento Química, Faculdade de Ciências e Tecnologia, Universidade Nova de Lisboa, Caparica, Portugal*

JULIEN MANDON • *Department of Molecular and Laser Physics, Radboud University Nijmegen, Nijmegen, The Netherlands*

SONAL MISHRA • *National Institute of Plant Genome Research, New Delhi, India*

JOSÉ J.G. MOURA • *UCIBIO, REQUIMTE, Departamento Química, Faculdade de Ciências e Tecnologia, Universidade Nova de Lisboa, Caparica, Portugal*

MARIN J. MUELLER • *Julius-von-Sachs Institute of Biosciences, University of Wuerzburg, Wuerzburg, Germany*

LUIS A.J. MUR • *Institute of Biological, Environmental, and Rural Sciences, Aberystwyth University, Wales, UK*

MAREK PETŘIVALSKÝ • *Department of Biochemistry, Faculty of Science, Palacky University, Olomouc, Czech Republic*

PRADEEP K. PATHAK • *National Institute of Plant Genome Research, New Delhi, India*

SUSANA PUNTARULO • *Physical Chemistry-Institute of Biochemistry and Molecular Medicine (IBIMOL), School of Pharmacy and Biochemistry, University of Buenos Aires-CONICET, Buenos Aires, Argentina*

AGEPATI S. RAGHAVENDRA • *Department of Plant Sciences, School of Life Sciences, University of Hyderabad, Hyderabad, India*

STEFAN RUEMER • *Julius-von-Sachs Institute of Biosciences, University of Wuerzburg, Wuerzburg, Germany*

YASUKO SAKIHAMA • *Research Faculty of Agriculture, Hokkaido University, Sapporo, Hokkaido, Japan*

BEATRIZ SÁNCHEZ-CALVO • *Group of Biochemistry and Cell Signaling in Nitric Oxide, Department of Biochemistry and Molecular Biology, University of Jaén, Jaén, Spain*

JEROME SANTOLINI • *Laboratoire Stress Oxydant et Détoxication, CNRS, Gif-sur-Yvette, France; iBiTec-S, CEA, Gif-sur-Yvette, France*

TEREZA TICHÁ • *Department of Biochemistry, Faculty of Science, Palacky University, Olomouc, Czech Republic*

RAQUEL VALDERRAMA • *Group of Biochemistry and Cell Signaling in Nitric Oxide, Department of Biochemistry and Molecular Biology, University of Jaén, Jaén, Spain*

ELODIE VANDELLE • *Dipartimento di Biotecnologie, Università degli Studi di Verona, Verona, Italy*

AAKANKSHA WANY • *National Institute of Plant Genome Research, New Delhi, India*

NAOKO S. WATANABE • *Faculty of Science, University of the Ryukyus, Nishihara, Okinawa, Japan*

DAVID WENDEHENNE • *Université de Bourgogne, UMR 1347 Agroécologie, Dijon, France; ERL CNRS 6300, Dijon, France*

HIDEO YAMASAKI • *Faculty of Science, University of the Ryukyus, Nishihara, Okinawa, Japan*

MANDA YU • *Institute of Molecular Plant Sciences, School of Biological Sciences, University of Edinburgh, Edinburgh, UK*

Chapter 1

An Overview of Methods in Plant Nitric Oxide (NO) Research: Why Do We Always Need to Use Multiple Methods?

Hideo Yamasaki, Naoko S. Watanabe, Yasuko Sakihama, and Michael F. Cohen

Abstract

The free radical nitric oxide (NO) is a universal signaling molecule among living organisms. To investigate versatile functions of NO in plants it is essential to analyze biologically produced NO with an appropriate method. Owing to the uniqueness of NO, plant researchers may encounter difficulties in applying methods that have been developed for mammalian study. Based on our experience, we present here a practical guide to NO measurement fitted to plant biology.

Key words Chemiluminescence detection, cPTIO, DAF, Electrochemical detection, Nitric oxide, RNS, ROS, RSS

1 Introduction

Scientific progress is made not only through discoveries and construction of new theories but also through the development of new technologies. Life science is not an exception. In a half-century, biochemistry and molecular biology have established many technologies and methodologies that enable us to investigate biomolecules including proteins, lipids, and nucleic acids. Nitric oxide (NO) exhibits unique characteristics contrary to these conventional biomolecules: it is simple, small, ubiquitous, and unstable [1]. However, despite the potential significance of NO in the physiological functions of plants [2] fundamental progress in this field has not matched our expectations [3–5]. The serious confusion regarding the identity of NO-producing enzymes in the 2000s is emblematic of the difficulties NO research has posed to the field of plant biology [3, 6]. Some confusion might be due to misinterpretation of results obtained by different techniques. Based on our

Kapuganti Jagadis Gupta (ed.), *Plant Nitric Oxide: Methods and Protocols*, Methods in Molecular Biology, vol. 1424, DOI 10.1007/978-1-4939-3600-7_1, © Springer Science+Business Media New York 2016

experience with NO measurements, we present here a practical guide on the choice of technique for measuring NO from plant materials.

2   To Know What You Measure

The NO molecule is commonly referred to as "nitric oxide" in the life sciences. According to systematic nomenclature, NO should be referred to as nitrogen monoxide, and it is advised to add a "dot" on the shoulder (NO•) because the molecule is a radical [7]. Following the tradition in life science, here we use "NO" and "nitric oxide" for simplification. To date, it has been confirmed that NO is produced in living organisms by two distinct routes, namely the arginine pathway and the nitrite pathway [1, 3, 8]. The arginine pathway includes the enzymatic synthesis of NO from L-arginine, oxygen, and NADPH by NO synthase (NOS), which requires the cofactors heme, FAD, FMN, and tetrahydrobiopterin [3]. The nitrite pathway involves multiple routes and mechanisms to produce NO from nitrite (NO_2^-) [8–10]:

$$L - arginine + 2O_2 + 1/2 NADPH \rightarrow L - citrulline + NO$$
$$+ 1/2 NADP^+ + 2H_2O$$

$$NO_{2^-} + H^+ + e^- \rightarrow NO + OH^-$$

In addition to NO, researchers should be reminded that there are two redox-related but chemically distinct species: nitroxyl anion (NO^-, oxonitrate (1–)) and nitrosonium cation (NO^+, nitrosyl cation):

$$NO + e^- \rightarrow NO^-$$

$$NO \rightarrow NO^+ + e^-$$

In the context of NO measurement, researchers need to be aware of these molecules. Sodium nitroprusside ($Na_2[Fe(CN)_5NO]$) known as SNP, which plant biologists often employ as a chemical NO donor (in spite of its simultaneous release of cyanide and ferrous iron), gives primarily NO^+, not NO, into solution [11, 12]. It is important to note that SNP does not spontaneously release NO but produces the molecule in the presence of reducing agents in the tissues under physiological conditions [13].

Contrasted with NO^+ formation, the reduction of NO to NO^- can easily occur in vivo through the action of biological reducing agents [7, 11]. Since interconversion of NO^-, NO, and NO^+ can occur under physiological conditions, researchers are recommended to take into account the diversity and complexity of chemistry of NO in designing experiments (including taking care in the selection of chemical NO donors) as well as interpreting results. In

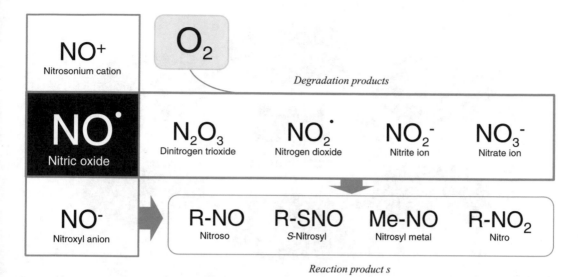

Fig. 1 NO and its derived nitrogen species. Major molecules that should be considered in NO measurement are presented. Note that each name follows the widely accepted one rather than the IUPAC systematic nomenclature

fact, NO^-, NO, and NO^+ display distinctive effects on living organisms [11]. Figure 1 illustrates major target molecules that should be considered in NO measurements.

3   Criteria for the Selection

For a new researcher who plans to measure NO in plant materials, it may be difficult to select a method from the number of options. Figure 2 represents six criteria that we consider important not only in the selection but also in the evaluation of results obtained with a single method. The most important criterion is specificity of a method for NO. This is probably a major cause for many of the confusions in NO research. The sensitivity is also crucial to the selection. Since NO is an air pollutant, a variety of instruments are commercially available to measure NO along with NO_2. Being designed for detecting an atmospheric high concentration of NO as ppm, most of those are not suitable for biological purposes. For life science, in general, a sensitivity detecting NO as ppb in air or nM in liquid is needed. The time scale you plan to measure is another important criterion [14]. The recognition of artifact risk is also necessary to avoid misleading to a wrong conclusion. In addition to these fundamental criteria, equipment availability as well as skill requirements are practical considerations in deciding upon a method. Therefore, in the selection of a method for NO measurement it is advisable to find the best balance between these multiple aspects.

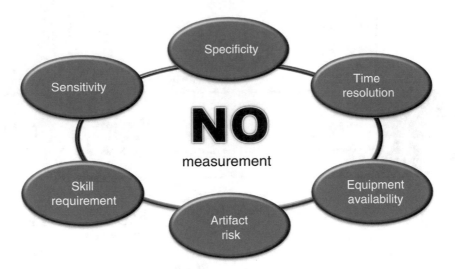

Fig. 2 Six criteria for NO measurement. New researchers may feel difficulty in selecting a method for NO measurement from a number of options. We suggest six important criteria for the selection of a method. It is advisable to find the best balance between these multiple aspects

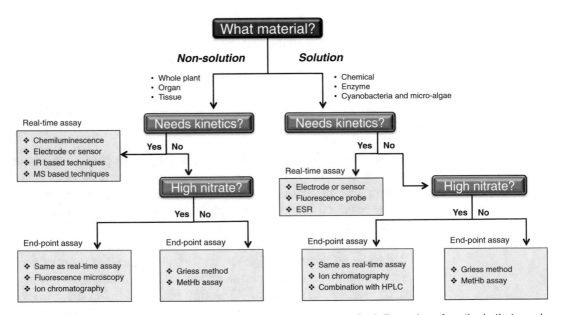

Fig. 3 A practical decision tree for selecting a NO measurement method. Examples of methods that can be carried out by most laboratories are shown. See other chapters for the latest techniques

Based on our experience, we present a decision tree for selecting an NO measurement suitable for plant biology and biochemistry (Fig. 3). The first question is whether you intend to measure NO in solution or non-solution. You may measure NO in a solution to study chemistry, enzymology, biochemistry, or algal physiology. To analyze a whole plant, organ, or tissue, you may follow the non-solution path of the chart. The second question is about the need of kinetics or time course. The last question is about the presence of interfering molecules that are abundantly contained in plant materials. Unlike animal tissues, green plants can accumulate high amounts of nitrate (NO_3^-). Nitrate and nitrite are potential substrates for NO production [15], and they give a high background to the Griess method. This is a reason why the method has not been widely applied in plant NO research (see below for details). Also, green plant tissues contain abundant antioxidants or reductants such as ascorbate and polyphenols [16] which may affect certain assays. The presence of photosynthetic pigments is also a unique potential risk in plants because they act as photosensitizers to produce oxidants including reactive oxygen species (ROS) under light [9].

4 Measurement of Gaseous NO

Measurement of gaseous NO is required to study long-range signal transmission mechanisms, including systemic responses in plant-plant, plant-microbe, or plant-herbivore interactions [17]. In general, experiments with gaseous molecules are technically much more difficult compared to those with solutions.

In mammalian research, measurements of gaseous NO are limited to special clinical areas, such as respiratory disorder studies. In plant science, however, biology and biochemistry of gas molecules including carbon dioxide (CO_2) and oxygen (O_2) have been extensively studied since they are the major substrate and product of photosynthesis [18]. In photosynthetic research, Clark-type O_2 electrodes are the first choice to measure O_2 evolution from a chloroplast suspension or from a leaf disk. To monitor CO_2 in the air, infrared (IR) absorption has been applied in biochemistry, physiology, and ecology. Except for the detection devices, the experimental setup and protocols for gaseous NO measurement are quite similar to those for O_2 and CO_2 measurements in photosynthesis studies. For measurement of gaseous NO, an IR absorption approach is also applicable with the absorption of NO at 5.3 μm in the infrared wavelength region [19]. Laser photoacoustic detection (LAPD) and quantum cascade lasers (QCL) that are based on IR absorption have been recently developed to quantify NO release from plants [19, 20]. Humidity (water vapor), which absorbs IR, is a strong interference factor for these IR-based technologies making a trapping system essential.

Mass spectroscopy (MS)-based techniques have been improved to fit into the demands of continuous NO measurements [21]. The specificity of MS-based techniques is promising but limited accessibility to the instruments is a major obstacle to their widespread use. From a practical point of view, we recommend plant biologists to consider the chemiluminescence technique for studying NO in air [22]. The device is commercially available as a medical instrument, originally designed for diagnosis of asthma through quantification of NO in exhaled air. The great advantage of this technique is its easy handling. Since the principle of the technique is to count photons emitted in the reaction between NO and ozone (O_3), confirmation may be necessary if plants release other volatile molecules that can react with O_3 to emit photons as similar to NO [23].

5 End-Point Assay

NO dissolves in aqueous solution with a solubility of 1.9 mM/atm [24]. In addition to the detection of gaseous NO, measurement of NO in a liquid sample is also required to explore signaling and regulatory mechanisms associated with NO. Application of electron spin resonance (ESR) or electron paramagnetic resonance (EPR) [25], electrochemical detection [26], fluorescence spectrophotometric [15], or fluorescence microscopic [27] techniques can be applied to measure NO in the liquid phase. If one cannot access these apparatuses, a colorimetric assay with an absorption spectrometer would be a practical option to estimate levels of NO produced. The Griess method and methemoglobin assay can be used to meet such demands.

The Griess method was originally designed to quantify NO_2^- or NO_3^- in a solution [28]. Stuehr and Marletta had applied the method to quantify NO produced in cells [29]. The Griess method is one of the most widely used NO detection procedures, especially in mammalian studies. The Griess method is also applicable for assessing gaseous NO with a CrO_3 oxidizer column; NO_2 converted from NO can be trapped in a Griess reagent solution and the gas degrades to NO_2^- that can be quantified spectrophotometrically without any special device or apparatus [30].

Because NO is not completely oxidized to NO_2^- or NO_3^- particularly under in vivo situations, the inaccuracy of this method is its major problem [31]. Furthermore, as an end-point-type procedure this method is not suitable for making real-time NO measurements. The methemoglobin assay, which quantifies NO through the conversion from oxyhemoglobin to methemoglobin spectrophotometrically, can also be categorized as an end-point assay. The assay includes potential problems similar to the Griess method [20]. In addition to specificity, the inclusion of colored molecules disturbs absorbance measurement so that a pretreatment to remove such molecules or pigments before the assay or combining with high-performance liquid chromatography (HPLC) is necessary. As a more accurate alternative to these methods we recommend use of ion chromatography HPLC for quantification of NO_2^- or NO_3^-.

6   Real-Time Measurement of NO

Signal transduction processes, in general, involve sequential events including a signal amplification process or cascade machinery as seen in Ca^{2+} signaling mechanisms [32, 33]. Supposing that NO is involved in plant signaling functions, one cannot assume that NO is constantly produced during the process. It is highly possible that NO may be produced only with a certain timing and at low levels. Such spiky NO emission was observed in early plant NO studies. In marine algae, abrupt NO emission was found just at the offset of illumination [34], apparently due to the shutting down of photosynthesis [27]. The observation could not have been made with an end-point assay mentioned above.

To explore a physiological response consisting of multiple steps and reactions, real-time measurement of NO is essential. Historically, photosynthetic electron transport in the chloroplasts has been characterized by kinetics studied at various time scales. Chlorophyll *a* fluorescence measurement, for example, has been investigated to monitor electron flow in the light reactions [35]. The induction and quenching sequence correspond to redox reactions along with subsequent ATP synthesis [36]. Plant biologists with experience in photosynthetic measurement have a background familiarizing them to such kinetic studies and real-time measurements.

Figure 4 shows practical examples for the real-time measurement of NO in liquid and gas phases. We used a Clark-type NO electrode to monitor nitrite-induced NO production in the unicellular alga *Chlamydomonas reinhardtii* [27]. The Clark electrode or probe was originally designed for the detection of oxygen. Oxidation of NO at the cathode, which is separated from the solution by electrolyte and a thin NO-selectively permeable membrane, generates a current that is directly proportional to the concentration of NO [26]. The experimental setup is quite similar to the measurement of photosynthetic O_2 evolution with a Clark-type O_2 electrode. To verify NO production, we used bovine hemoglobin to eliminate the signal [15, 27]. Since most of the NO scavenging actions involve oxidoreduction, we avoided the use of chemical NO scavengers in electrochemical detection with an electrode. Although hemoglobin is heavily colored, it fundamentally does not affect the electrochemical detection unlike in absorption spectrophotometric measurement. NO simply binds to the hemes without potentially confounding reactions.

It should be noted that Clark-type electrodes are very sensitive to temperature changes. Even small temperature changes due to enthalpy of fusion (the heat of fusion) can be detected as a signal. Therefore, to carry out NO measurement with the electrode, careful control of temperature is strongly recommended. A water-jacketed reaction chamber is a prerequisite to measuring NO by this method.

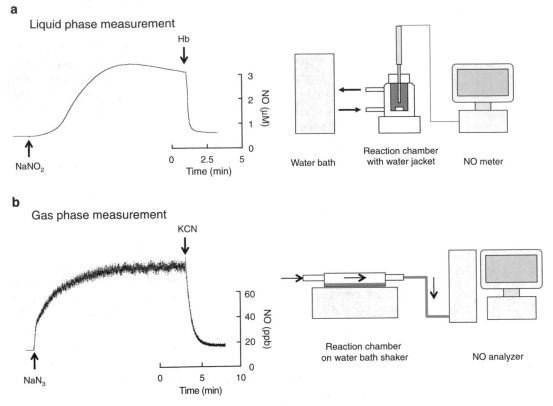

Fig. 4 Real-time measurement of NO with plant materials. (**a**) Liquid-phase NO measurement with an NO electrode. Time course of nitrite (NO_2^-)-induced NO production in the unicellular alga *Chlamydomonas reinhardtii* is shown on the *left*. With a time lag after the addition of nitrite, NO is produced. The signal is eliminated by hemoglobin. (**b**) Gas-phase NO measurement with a chemiluminescence detector. Time course of azide (N_3^-)-induced NO production in the water fern *Azolla pinnata*. KCN abolishes the signal from the fern. The *graphs* are redrawn after Sakihama et al. [27] and Gurung et al. [37] for (**a**) and (**b**), respectively

The chemiluminescence technique also gives real-time NO monitoring. This technique is useful for monitoring gaseous NO emission from whole plants [22]. In contrast with electrochemical detection using NO electrodes, the chemiluminescence technique is easy in handling; it is a recommended method for many purposes in plant biology. In the case of the tiny water fern *Azolla*, the effects of exogenously added chemicals can be monitored similar to liquid measurements [37]. In Fig. 4b, KCN was added to distinguish enzymatic NO synthesis from nonenzymatic or chemical NO production.

7   Why Multiple Techniques Are Required: Case Studies

Whatever the method chosen, researchers must be aware of potential errors that can be avoided by knowing the principles behind them. There is no perfect method in experimental

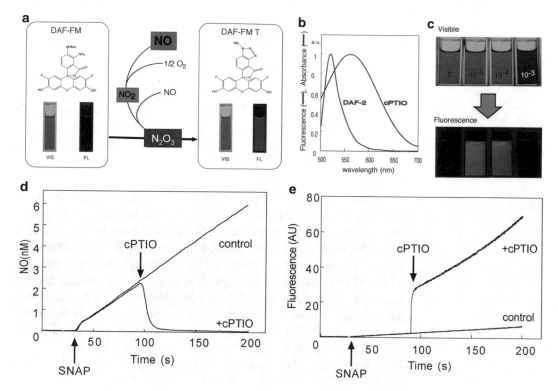

Fig. 5 DAF and cPTIO. (**a**) Schematic overview of fluorometric NO detection with DAFs. (**b**) Fluorescence and absorption spectra of DAF and cPTIO, respectively. Visible and fluorescence images of DAF solutions contained cPTIO at different concentrations (0, 10^{-5}, 10^{-4}, and 10^{-3} M). Black light was applied for visual observation of DAF fluorescence. Effects of cPTIO on NO level (**d**) and DAF fluorescence (**e**). NO concentration and DAF fluorescence were monitored with a NO-specific electrode (**d**) and fluorescence spectrophotometer (**e**), respectively. cPTIO (50 μM) was added to the 10 μM DAF-2 solution (pH 7.4 phosphate buffer) at the *arrows* indicated. NO was chemically generated with SNAP (250 μM). The *graphs* are redrawn after Arita et al. [39]

science; awareness of application limits is always needed to avoid misinterpretation of data. This is particularly true for the detection of NO in living systems. Therefore, we strongly recommend employment of at least two methods having distinct detection principles [38, 39].

In plant science, cPTIO (2-(4-carboxyphenyl)-4,4,5,5-tetramethylimidazoline-1-oxyl-3-oxide) has been widely used to establish the involvement of NO in physiological phenomena. It should be noted that cPTIO is an organic radical that was used as a spin-trapping agent to detect NO in ESR measurement [40]. Since the reaction is stoichiometric and specific to NO, Akaike and co-workers applied cPTIO for removing NO from a solution [25].

Figure 5 demonstrates the effect of cPTIO on fluorescence intensity of diaminofluorescein-2 triazole (DAF-2T) that is formed from diaminofluorescein-2 (DAF-2). It is indeed true that cPTIO abolishes the NO signal monitored with an NO electrode

(Fig. 5d), a result confirming that cPTIO is a potent NO scavenger, oxidizing NO to form NO_2 via the intermediate N_2O_3:

$$cPTIO + NO \rightarrow cPTI + NO_2$$

Contrary to a widely held presumption, the NO scavenger cPTIO does not suppress but actually enhances the conversion of DAF-2 into the DAF-2T fluorophore (Fig. 5e). This observation is explained by the fact that DAFs do not directly react with NO but with its oxidized products (such as N_2O_3 or NO^+ [20, 39]) that are formed in the presence of O_2.

For most plant biologists it may be difficult to follow the actual reaction chemistry and fluorescence quantum yield underlying a phenomenon. In plant science, a similar situation can be found in the application of pulse amplitude modulation (PAM) chlorophyll *a* fluorescence measurement [41]. Although the mechanisms involved in quenching chlorophyll *a* fluorescence remain controversial, the apparatus is welcomed by plant ecological researchers to quantify a number of parameters in photosynthesis including quantum yield (Fv/Fm) and electron transport rate (ETR) [42, 43]. Similar to the PAM fluorescence imaging technique [44], DAF fluorescence is a powerful tool for visualization of a local NO production [27]. As many researchers have pointed out, there is an application limit of the DAF series [38, 39, 45]. To reduce the risk of detecting artifacts, therefore, one should not deduce a conclusion solely from the results obtained with a single technique. This is particularly true in emerging new fields such as NO research. Confirmation with multiple methods is ideal [38, 39].

8   Assembling the Pieces of a Jigsaw Puzzle

The ROS superoxide (O_2^-) that is produced in cells under stress conditions reacts with NO to produce the RNS peroxynitrite (ONOO⁻):

$$NO + O_2^- \rightarrow ONOO^-$$

The reaction occurs much faster than dismutation of O_2^- by the enzyme superoxide dismutase (SOD) [9]. This reaction likely happens in whole plants under stress conditions. If ROS and NO are simultaneously produced under such conditions, NO may not be detected outside of the cells, a potential factor that can lead to an underestimation of NO production. Another key player is thiols (-SH) contained in living organisms. NO can exist in the cell as nitrosothiols that serve as an endogenous NO donors or reservoirs. *S*-nitrosoglutathione (GSNO) is known not only in the context of physiological relevance but also as an experimental NO donor [46]. Nitrosylation occurs also with metals to form nitrosyl complexes such as Fe-NO on heme. Thus, the nitrosylation reaction, as well as nitrosation and nitration reactions, could cause underestimation of NO production when it is monitored outside of the cells.

From a practical aspect of NO measurement, we advise researchers to pay much attention to oxygen tension in an experimental system. This is because (1) the degradation of NO proceeds in the presence of O_2, (2) ROS which reacts with NO is produced with O_2, and (3) instead of O_2, under anoxic conditions electron transport systems, such as in mitochondria, may reduce NO_2^- as an electron acceptor to form NO [47]. Therefore, we consider that O_2 is a key determinant for evaluating NO chemical biology. Although anoxic conditions are rarely found in plants except for roots in soils, the exhausting of O_2 in a closed vessel frequently happens in laboratory experiments with living cells or plants. In this context, for ESR experiments using a capillary tube or microscopic observation with a cover glass, special care should be taken if the sample consumes O_2 through respiration activity. In the case of experiments using DAFs one should also be reminded that O_2 is required to form the fluorescent DAF-2T (Fig. 5a).

To date, there have been a number of studies that measure biologically produced NO with diverse techniques and materials (enzymes, organelles, cells, tissues, organs, and whole plants). To draw a whole picture of the physiological function of NO in plants, the integration of results is required. However, results from different methods may sometimes conflict. Some such conflicts have their origin in the failure to have properly considered detection principles in designing an experiment; the shape of each piece of the jigsaw puzzle may not have been formatted to assemble a whole picture.

9 Future Prospects

Plant NO research in its modern style has a 20-year history [5]. In contrast with mammalian NO research, however, fundamental progress of the field, especially on mechanistic aspects, has not matched our expectations [3]. Without a solid framework, we have not yet reached a point at which we can readily translate the outcome of plant NO research to applied sciences such as agriculture [17]. Obviously, difficulty in NO measurement is one of the major problems. The complexity of the chemistry also impedes more rapid progress for NO plant research. To overcome the difficulties, the development of more user-friendly NO-measuring devices is required. Such devices should be easy to handle, not be expensive, have high specificity, and ideally be applicable for real-time measurement (Fig. 2). Looking at the history of pH measurement (from glass electrode to ISFET sensor) or spectroscopic assays (from photomultiplier to photodiode array), it is possible to expect electronic NO sensors in the future that meet the six criteria with a good balance.

Recently, hydrogen sulfide (H_2S) has been suggested to display signaling functions similar to NO in mammals [48]. H_2S gives reactive sulfur species (RSS) that can be distinguished from ROS and RNS [49]. It is interesting to note that these ROS (O_2^-), RNS (NO), and RSS (H_2S) all appear to be produced by chloroplasts in the course of photosynthesis (CO_2 assimilation), nitrate assimilation, and sulfate assimilation metabolisms, presumably as by-products. As seen in the S-nitrosyl formation with thiols (e.g., GSNO), the interaction between RNS and RSS is physiologically important. Although chemical biology needs to wait for the development of new technology to measure biologically produced RSS, our final goal will be to understand the dynamics between ROS, RNS, and RSS in living plants. Until the delivery of the big piece of the puzzle (RSS), we have to carefully investigate plant NO functions with multiple verifications.

Acknowledgements

Due to space limitations we were not able to cite many brilliant works on plant NO research that have applied the methods described in this chapter. Please refer to other chapters for such investigations. We thank Dr. Jon Fukuto for his valuable comments on this chapter. This work was supported by the grants to H.Y. from the Japanese Ministry of Education, Science, Culture and Sports.

References

1. Yamasaki H (2005) The NO world for plants: achieving balance in an open system. Plant Cell Environ 28:78–84

2. Mur LAJ, Mandon J, Persijn S et al (2013) Nitric oxide in plants: an assessment of the current state of knowledge. AoB Plants 5:1–17

3. Yamasaki H, Itoh RD, Bouchard JN et al (2011) Nitric oxide synthase-like activities in plants. In: Foyer CH, Zhang H (eds) Nitrogen metabolism in plants in the post-genomic era, vol 42, Annual Plant Reviews. Blackwell Publishing Ltd, West Sussex, pp 103–125

4. Leshem YY (2000) Nitric oxide in plants: occurrence, function and use. Kluwer Academic Publishers, Dordrecht

5. Yamasaki H (2004) Nitric oxide research in plant biology: its past and future. In: Magalhaes JR, Singh RP, Passos LP (eds) Nitric oxide signaling in higher plants. Focus on plant molecular biology. Studium Press, Houston, pp 1–23

6. Yamasaki H, Cohen MF (2006) NO signal at the crossroads: polyamine-induced nitric oxide synthesis in plants? Trends Plant Sci 11:522–524

7. Koppenol WH, Traynham JG (1996) Say NO to nitric oxide: nomenclature for nitrogen- and oxygen-containing compounds. Methods Enzymol 268:3–7

8. Yamasaki H, Watanabe NS, Fukuto J et al (2014) Nitrite-dependent nitric oxide production pathway: diversity of NO production systems. In: Tsukahara H, Kaneko K (eds) Studies on pediatric disorders. Springer, New York, pp 35–54

9. Yamasaki H (2000) Nitrite-dependent nitric oxide production pathway: implications for involvement of active nitrogen species in photoinhibition *in vivo*. Philos Trans R Soc Lond B Biol Sci 355:1477–1488

10. Yamasaki H, Sakihama Y, Takahashi S (1999) An alternative pathway for nitric oxide produc-

tion in plants: new features of an old enzyme. Trends Plant Sci 4:128–129

11. Hughes MN (1999) Relationships between nitric oxide, nitroxyl ion, nitrosonium cation and peroxynitrite. Biochim Biophys Acta 1411:263–272

12. Rogstam A, Larsson JT, Kjelgaard P et al (2007) Mechanisms of adaptation to nitrosative stress in *Bacillus subtilis*. J Bacteriol 189:3063–3071

13. Bates JN, Baker MT, Guerra R et al (1991) Nitric oxide generation from nitroprusside by vascular tissue: evidence that reduction of the nitroprusside anion and cyanide loss are required. Biochem Pharmacol 42:S157–S165

14. Liu XP, Liu QH, Gupta E et al (2005) Quantitative measurements of NO reaction kinetics with a Clark-type electrode. Nitric Oxide 13:68–77

15. Yamasaki H, Sakihama Y (2000) Simultaneous production of nitric oxide and peroxynitrite by plant nitrate reductase: *in vitro* evidence for the NR-dependent formation of active nitrogen species. FEBS Lett 468:89–92

16. Sakihama Y, Cohen MF, Grace SC et al (2002) Plant phenolic antioxidant and prooxidant activities: phenolics-induced oxidative damage mediated by metals in plants. Toxicology 177:67–80

17. Cohen MF, Mazzola M, Yamasaki H (2006) Nitric oxide research in agriculture: bridging the plant and bacterial realms. In: Rai K, Takabe T (eds) Abiotic stress tolerance in plants. Springer, Dordrecht, pp 71–90

18. Takahashi S, Tamashiro A, Sakihama Y et al (2002) High-susceptibility of photosynthesis to photoinhibition in the tropical plant *Ficus microcarpa* L. f. cv. Golden Leaves. BMC Plant Biol 2:1–8

19. Mur LAJ, Mandon J, Cristescu SM et al (2011) Methods of nitric oxide detection in plants: a commentary. Plant Sci 181:509–519

20. Mur LAJ, Santosa IE, Laarhoven LJJ et al (2005) Laser photoacoustic detection allows in planta detection of nitric oxide in tobacco following challenge with avirulent and virulent *Pseudomonas syringae* pathovars. Plant Physiol 138:1247–1258

21. Conrath U, Amoroso G, Kohle H et al (2004) Non-invasive online detection of nitric oxide from plants and some other organisms by mass spectrometry. Plant J 38:1015–1022

22. Rockel P, Strube F, Rockel A et al (2002) Regulation of nitric oxide (NO) production by plant nitrate reductase *in vivo* and *in vitro*. J Exp Bot 53:103–110

23. Hossain KK, Itoh RD, Yoshimura G et al (2010) Effects of nitric oxide scavengers on thermoinhibition of seed germination in *Arabidopsis thaliana*. Russ J Plant Physiol 57:222–232

24. Wink DA, Grisham MB, Mitchell JB et al (1996) Direct and indirect effects of nitric oxide in chemical reactions relevant to biology. Methods Enzymol 268:12–31

25. Akaike T, Maeda H (1996) Quantitation of nitric oxide using 2-phenyl-4,4,5,5-tetramethylimidazoline-1-oxyl 3-oxide (PTIO). Methods Enzymol 268:211–221

26. Malinski T, Mesaros S, Tomboulian P (1996) Nitric oxide measurement using electrochemical methods. Methods Enzymol 268:58–69

27. Sakihama Y, Nakamura S, Yamasaki H (2002) Nitric oxide production mediated by nitrate reductase in the green alga *Chlamydomonas reinhardtii*: an alternative NO production pathway in photosynthetic organisms. Plant Cell Physiol 43:290–297

28. Griess P (1879) Bemerkungen zu der Abhandlung der HH. Weselsky und Benedikt. Ueber einige Azoverbindungen. Ber Dtsch Chem Ges 12:426–428

29. Stuehr DJ, Marletta MA (1985) Mammalian nitrate biosynthesis: mouse macrophages produce nitrite and nitrate in response to *Escherichia coli* lipopolysaccharide. PNAS 82:7738–7742

30. Vitecek J, Reinohl V, Jones RL (2008) Measuring NO production by plant tissues and suspension cultured cells. Mol Plant 1:270–284

31. Hunter RA, Storm WL, Coneski PN et al (2013) Inaccuracies of nitric oxide measurement methods in biological media. Anal Chem 85:1957–1963

32. Berridge MJ, Bootman MD, Roderick HL (2003) Calcium signalling: dynamics, homeostasis and remodelling. Nat Rev Mol Cell Biol 4:517–529

33. Berridge MJ, Lipp P, Bootman MD (2000) The versatility and universality of calcium signalling. Nat Rev Mol Cell Biol 1:11–21

34. Mallick N, Rai LC, Mohn FH et al (1999) Studies on nitric oxide (NO) formation by the green alga *Scenedesmus obliquus* and the diazotrophic cyanobacterium *Anabaena doliolum*. Chemosphere 39:1601–1610

35. Maxwell K, Johnson GN (2000) Chlorophyll fluorescence - a practical guide. J Exp Bot 51:659–668

36. Gilmore AM, Yamasaki H (1998) 9-aminoacridine and dibucaine exhibit competitive interactions and complicated inhibitory effects that interfere with measurements of Δ pH and xanthophyll cycle-dependent photosystem II energy dissipation. Photosynth Res 57:159–174

37. Gurung S, Cohen MF, Yamasaki H (2014) Azide-dependent nitric oxide emission from the water fern *Azolla pinnata*. Russ J Plant Physiol 61:543–547

38. Gupta KJ, Igamberdiev AU (2013) Recommendations of using at least two different methods for measuring NO. Front Plant Sci 4:1–4

39. Arita NO, Cohen MF, Tokuda G et al (2007) Fluorometric detection of nitric oxide with diaminofluoresceins (DAFs): applications and limitations for plant NO research. In: Lamattina L, Polacco J (eds) Nitric oxide in plant growth, development and stress physiology. Springer, Würzburg, pp 269–280

40. Az-ma T, Fujii K, Yuge O (1994) Reaction between imidazolineoxil N-oxide (carboxy-PTIO) and nitric oxide released from cultured endothelial cells: quantitative measurement of nitric oxide by ESR spectrometry. Life Sci 54:185–190

41. Lichtenthaler HK, Buschmann C, Knapp M (2005) How to correctly determine the different chlorophyll fluorescence parameters and the chlorophyll fluorescence decrease ratio R_{Fd} of leaves with the PAM fluorometer. Photosynthetica 43:379–393

42. Ralph PJ, Gademann R (2005) Rapid light curves: a powerful tool to assess photosynthetic activity. Aquat Bot 82:222–237

43. Takahashi S, Nakamura T, Sakamizu M et al (2004) Repair machinery of symbiotic photosynthesis as the primary target of heat stress for reef-building corals. Plant Cell Physiol 45:251–255

44. Hossain KK, Nakamura T, Yamasaki H (2011) Effect of nitric oxide on leaf non-photochemical quenching of fluorescence under heat-stress conditions. Russ J Plant Physiol 58:629–633

45. Planchet E, Kaiser WM (2006) Nitric oxide (NO) detection by DAF fluorescence and chemiluminescence: a comparison using abiotic and biotic NO sources. J Exp Bot 57:3043–3055

46. Singh RJ, Hogg N, Joseph J et al (1996) Mechanism of nitric oxide release from *S*-nitrosothiols. J Biol Chem 271:18596–18603

47. Planchet E, Gupta KJ, Sonoda M et al (2005) Nitric oxide emission from tobacco leaves and cell suspensions: rate limiting factors and evidence for the involvement of mitochondrial electron transport. Plant J 41:732–743

48. Ono K, Akaike T, Sawa T et al (2014) Redox chemistry and chemical biology of H_2S, hydropersulfides, and derived species: Implications of their possible biological activity and utility. Free Radic Biol Med 77:82–94

49. Gruhlke MCH, Slusarenko AJ (2012) The biology of reactive sulfur species (RSS). Plant Physiol Biochem 59:98–107

Chapter 2

Chemiluminescence Detection of Nitric Oxide from Roots, Leaves, and Root Mitochondria

Aakanksha Wany, Alok Kumar Gupta, Aprajita Kumari, Shika Gupta, Sonal Mishra, Ritika Jaintu, Pradeep K. Pathak, and Kapuganti Jagadis Gupta

Abstract

NO is a free radical with short half-life and high reactivity; due to its physiochemical properties it is very difficult to detect the concentrations precisely. Chemiluminescence is one of the robust methods to quantify NO. Detection of NO by this method is based on reaction of nitric oxide with ozone which leads to emission of light and amount of light is proportional to NO. By this method NO can be measured in the range of pico moles to nano moles range. Using direct chemiluminescence method, NO emitted into the gas stream can be detected whereas using indirect chemiluminescence oxidized forms of NO can also be detected. We detected NO using purified nitrate reductase, mitochondria, cell suspensions, and roots; detail measurement method is described here.

Key words Chemiluminescence, Mitochondria, Nitric oxide, Cell suspensions, Hydroponics

1 Introduction

Nitric oxide is a free radical molecule that plays an important role in plant growth development and stress [1]. In recent years nitric oxide roles in plants have been emerging. Various enzymes have been characterized for their ability to generate NO [2, 3]: cytosolic nitrate reductase (NR), mitochondria nitrite NO reductase, the plasma membrane nitrite: NO reductase (PM-NiNOR), xanthine oxidoreductase, NO synthase-like enzyme, polyamine (PA), and hydroxylamine (HA)-mediated pathways. In order to interpret NO roles, it is vital to know how much NO is produced from each enzyme system and its spatial and temporal generation pattern is also very important. Despite extensive research, accurate method to measure NO concentration is currently unavailable.

NO stability depends on its concentration. For instance, at 20 μM concentration NO has a half-life approximately 160 s whereas at 100 μM concentration half-life of NO is about 8 s [4]. NO

Kapuganti Jagadis Gupta (ed.), *Plant Nitric Oxide: Methods and Protocols*, Methods in Molecular Biology, vol. 1424, DOI 10.1007/978-1-4939-3600-7_2, © Springer Science+Business Media New York 2016

concentration varies according to specific stress and developmental stage. For instance under anoxia roots produce very high amount of NO continuously whereas during development cells produce minute amount of NO; therefore methods should be sensitive enough to measure NO at low and higher concentrations. In some cases NO production restricted to specific cells for instance during stomatal closing NO production increases in guard cells [5].

There are various methods that have been developed such as diamino fluorescence diacetate (DAF) method, chemiluminescence, electron spin resonance (ESR), mass spectrometry, Griess reagent assay, oxyhemoglobin assay, NO electrodes, laser-based infrared spectrometry (LAPD), and arginine-to-citrulline assay [6, 7, 31]. Each method has some advantages and disadvantages; for instance chemiluminescence is the most widely accepted method for measurement of NO; it can measure NO from pico moles to nano moles range (Table 1) but the disadvantage is that it can measure only NO emitted into gas stream [17]. The scavenged NO cannot be detected by this method. For detection of scavenged NO indirect chemiluminescence method is needed [18]. Another method called as laser-based photoacoustic detection of NO can measure NO very precisely but disadvantage is similar to chemiluminescence that it can measure only NO emitted into gas phase. The membrane inlet mass spectrometry (MIMS) is another method where NO emission can be measured very precisely, but the disadvantage is high cost and maintenance of the device; moreover NO that is emitted into gas phase only can be measured [19]. Another method to measure NO is using NO electrode [20, 21]. The advantage of this method is that NO can be measured in extract and cell suspensions treated with various elicitors or any abiotic stress. Disadvantage is that it can measure NO in gas phase only [22]. Electron spin resonance (ESR) is another method to measure NO in pico molar range. The advantage is that it measures NO very precisely, and the disadvantage is that it can measure NO that is in oxidized forms and measurement of emitted NO is not possible by this method [23]. Oxyhemoglobin assay is another method in which oxyhemoglobin (HbO_2) reaction with NO leads to production of methemoglobin (MetHb) and nitrate (NO^-_3). This method can measure NO in nano molar range [24]. The advantage is that it can be easily done but the disadvantage is that it can measure only trapped NO; NO emitted into gas phase is not possible to detect [25]; another disadvantage is that reactive oxygen species and low pH can interfere with measurement.

By using direct and indirect chemiluminescence one can measure both oxidized and reduced forms of NO.

The measurement of NO by chemiluminescence is usually performed by a NO analyzer (CLD 770 AL ppt, Eco-Physics, Switzerland). The principle for measuring NO is the measurement

Table 1
Chemiluminescence-based detection of NO in plants

S. No.	Plant type	Tissue	NO emission rates	Condition	
1	*Nicotiana benthamiana*	Excised leaves	50 nmol/gFW/h	Nitrate supplementation	[8]
2	*Hordeum vulgare*	Aleurone layers	15 nmol/gFW/h 0.025 nmol/gFW/h	Nitrite supplementation Aerobic conditions	
3	*Arabidopsis thaliana*	Leaves	0.4 nmol/gFW/h	Endogenous production	
4	*Nicotiana tabacum*	Suspension culture cells	0.38 nmol/gFW/h 0.79 nmol/gFW/h	Water treatment Cryptogein treatment	
5	*Helianthus annuus*	Leaves	170 nmol/gFW/h	Anoxia	[9]
6	*Glycine max*	Leaves	10 μmol/gFW/h	Nitrate treatment	[10, 11]
7	*Nicotiana tabacum*	Leaves	40 nmol/gFW/h	Pathogen infection	[1]
		Roots	9 nmol/gFW/h	Aerobic conditions	[12]
		WT Root mitochondria	<1 nmol/mg protein/h 3.5 nmol/mg protein/h <1 nmol/mg protein/h <1 nmol/mg protein/h	0.5 mM Nitrite/1 mM NADH/air 0.5 mM Nitrite/1 mM NADH/anoxia 0.5 mM Nitrite/1 mM NADH/20 μM myxothiazol 0.5 mM Nitrite/1 mM NADH/2.5 mM SHAM	[13]
		WT Root	<1 nmol/gFW/h 12 nmol/gFW/h 17 nmol/gFW/h	0.5 mM Nitrite/100 % air 0.5 mM Nitrite/0.025 % air 0.5 mM Nitrite/anoxia	
		WT Leaf mitochondria	<1 nmol/gFW/h 70 nmol/gFW/h	0.5 mM Nitrite/air 0.5 mM Nitrite/anoxia	

(continued)

Table 1
(continued)

S. No.	Plant type	Tissue	NO emission rates	Condition	
		NR-free *nia* 30 double mutant	<1 nmol/gFW/h	Nitrite/air	
			<1 nmol/gFW/h	Nitrite/anoxia	
		Root segments	<1 nmol/mg protein/h	0.5 mM Nitrite/anoxia	
			7 nmol/mg protein/h	Ammonia/tungstate/anoxia	
			3 nmol/gFW/h	0.5 mM Nitrite/air	
			40 nmol/gFW/h	0.5 mM Nitrite/anoxia	
			10 nmol/gFW/h	0.5 mM Nitrite/50 µM Myxothiazol/anoxia	
			<1 nmol/gFW/h	0.5 mM Nitrite/2.5 mM SHAM/anoxia	
		NR-free *nia* double mutant	4 nmol/mg protein/h	0.5 mM Nitrite/1 mM NADH/anoxia	
			1.5 nmol/mg protein/h	0.5 mM Nitrite/1 mM NADH/20 µM Myxothiazol	
		Root mitochondria	<1 nmol/mg protein/h	0.5 mM Nitrite/1 mM NADH/2.5 mM SHAM	
		NR-free *nia* 30	<1 nmol/gFW/h	0.5 mM Nitrite/air	
		Leaf mitochondria	0 nmol/gFW/h	0.5 mM Nitrite/anoxia	
		WT	<1 nmol/gFW/h	Nitrite/air	
		Root segments	9 nmol/gFW/h	Nitrite/anoxia	
			<1 nmol/mg protein/h	0.5 mM Nitrite/anoxia	
			6 nmol/mg protein/h	Ammonia/tungstate/anoxia	
			<1 nmol/gFW/h	0.5 mM Nitrite/air	
			40 nmol/gFW/h	0.5 mM Nitrite/anoxia	
			10 nmol/gFW/h	0.5 mM Nitrite/50 µM Myxothiazol/anoxia	
			10 nmol/gFW/h	0.5 mM Nitrite/2.5 mM SHAM/anoxia	
			<1 nmol/gFW/h	0.5 mM Nitrite/1.5 mM KCN	
8	*Arabidopsis thaliana*	Leaves	80 pmol/gFW/h	*Pseudomonas* infection	[14]
9	*Pisum sativum*	Roots	14.4 ± 0.05 nmol/gFW/h	NOS activity	[15]
		Stems	37.8 ± 0.22 nmol/gFW/h		
		Leaves	7.2 ± 0.02 nmol/gFW/h		

No	Species	Sample	Value	Condition	Ref
10	Nicotiana tabacum	Leaves with petioles	0.01–0.05 nmol/gFW/h	Nitrate solution/Dark-light/air	[16]
		Leaves	100 nmol/gFW/h	Nitrogen flushing/air- Dark transient	
		Cell suspensions	0.01–0.05 nmol/gFW/h	Air- nitrogen transition	
			9.5–10 nmol/gFW/h	Dark/Anoxia	
			0.01–0.5 nmol/gFW/h	Reoxygenation	
11	NiR deficient Nicotiana tabacum mutant "clone 271"	Leaves	1–23 nmol/gFW/h	Dark-Light–dark transient	
			5–16 nmol/gFW/h	Dark/Anoxia	
12	Nicotiana tabacum cv. Xanthi and cv. Gatersleben	Cell suspensions	0.83 ± 0.31 nmol/gFW/h	Ammonium medium/air + nitrite addition	
			35.76 ± 13.01 nmol/gFW/h	Ammonium medium/Anoxia + nitrite addition	
			0.75 ± 0.31 nmol/gFW/h	Ammonium medium/air + 10 μM myxothiazol	
			0.36 ± 0.10 nmol/gFW/h	Ammonium medium/air + 2.5 mM SHAM	
			0.28 ± 0.10 nmol/gFW/h	Ammonium medium/air + 2 mM KCN	
			0.07 ± 0.03 nmol/gFW/h	Ammonium medium/air + 10 μM myxothiazol + 2.5 mM SHAM + 2 mM KCN	
			28.62 ± 10.29 nmol/gFW/h	Ammonium medium/Anoxia + 10 μM myxothiazol	
			22.66 ± 7.46 nmol/gFW/h	Ammonium medium/Anoxia + 2.5 mM SHAM	
			15.52 ± 6.71 nmol/gFW/h	Ammonium medium/Anoxia + 2 mM KCN	
			0.81 ± 0.52 nmol/gFW/h	Ammonium medium/Anoxia + 10 μM myxothiazol + 2.5 mM SHAM + 2 mM KCN	
			0.57 ± 0.09 nmol/gFW/h	Nitrate medium/air + nitrite addition	
			0.53 ± 0.09 nmol/gFW/h	Nitrate medium/air + 10 μM myxothiazol	
			0.37 ± 0.14 nmol/gFW/h	Nitrate medium/air + 2.5 mM SHAM	

(continued)

Table 1
(continued)

S. No.	Plant type	Tissue	NO emission rates	Condition
			0.32 ± 0.07 nmol/gFW/h	Nitrate medium/air + 2 mM KCN
			0.05 ± 0.02 nmol/gFW/h	Nitrate medium/air + 10 μM myxothiazol + 2.5 mM SHAM + 2 mM KCN
			24.05 ± 4.68 nmol/gFW/h	Nitrate medium/anoxia + nitrite addition
			28.83 ± 9.40 nmol/gFW/h	Nitrate medium/anoxia + 10 μM myxothiazol
			28.37 ± 5.47 nmol/gFW/h	Nitrate medium/anoxia + 2.5 mM SHAM
			30.73 ± 9.23 nmol/gFW/h	Nitrate medium/anoxia + 2 mM KCN
			1.35 ± 0.72 nmol/gFW/h	Nitrate medium/anoxia + 10 μM myxothiazol + 2.5 mM SHAM + 2 mM KCN
13	*Nicotiana tabacum* cv. *Xanthi*	Cell suspensions	$0.01–0.05$ nmol/gFW/h	Nitrate medium + water
			$0.2–0.8$ nmol/gFW/h	Nitrate medium + cryptogein
			$0.4–1.0$ nmol/gFW/h	Nitrate medium + cryptogein + NOS inhibitor
			$0.2–0.6$ nmol/gFW/h	Ammonium medium + cryptogein
			0 nmol/gFW/h	Nitrite + cPTIO/air
			$0.1–22$ nmol/gFW/h	Nitrite + cPTIO/anoxia
			$0.1–18.97$ nmol/gFW/h	Nitrite + cPTI/anoxia
			$0.1–0.2$ nmol/gFW/h	Nitrite + cryptogein
			$0.0–0.2$ nmol/gFW/h	Nitrite + cryptogein + cPTIO
			$0.1–0.6$ nmol/gFW/h	Nitrite + cryptogein + cPTI

[16]

14	*Pisum sativum*	Leaf slices	0 nmol/gFW/h	0.5 mM Nitrite/air
			80 nmol/gFW/h	0.5 mM Nitrite/nitrate/anoxia
			<1 nmol/gFW/h	0.5 mM Nitrite/ammonia/anoxia
		Leaf mitochondria	<1 nmol/mg Protein/h	0.5 mM Nitrite/1 mM NADH/air
			<1 nmol/mg Protein/h	0.5 mM Nitrite/1 mM NADH/anoxia
		Root segments	<1 nmol/gFW/h	0.5 mM Nitrite/air
			24 nmol/gFW/h	0.5 mM Nitrite/anoxia
		Purified root mitochondria	<1 nmol/mg Protein/h	0.5 mM Nitrite/1 mM NADH/air
			4.5 nmol/mg Protein/h	0.5 mM Nitrite/1 mM NADH/anoxia
			2 nmol/mg Protein/h	0.5 mM Nitrite/1 mM NADH/20 µM myxothiazol/anoxia
			1 nmol/mg Protein/h	0.5 mM Nitrite/1 mM NADH/2.5 mM SHAM/anoxia	[13]
15	*Hordeum vulgare*	Leaf slices	0 nmol/gFW/h	0.5 mM Nitrite/air
			130 nmol/gFW/h	0.5 mM Nitrite/nitrate/anoxia
			<1 nmol/gFW/h	0.5 mM Nitrite/ammonia/anoxia
		Leaf mitochondria	<1 nmol/mg Protein/h	0.5 mM Nitrite/1 mM NADH/air
			<1 nmol/mg Protein/h	0.5 mM Nitrite/1 mM NADH/anoxia
		Root segments	<1 nmol/gFW/h	0.5 mM Nitrite/air
			9 nmol/gFW/h	0.5 mM Nitrite/anoxia
		Purified root mitochondria	<1 nmol/mg Protein/h	0.5 mM Nitrite/1 mM NADH/air
			15 nmol/mg Protein/h	0.5 mM Nitrite/1 mM NADH/anoxia
			6 nmol/mg Protein/h	0.5 mM Nitrite/1 mM NADH/20 µM myxothiazol/anoxia
			3 nmol/mg Protein/h	0.5 mM Nitrite/1 mM NADH/2.5 mM SHAM/anoxia

of light emission resulting from the reaction between nitric oxide (NO) and ozone (O). In this method NO is measured directly. The reactions are described as follows:

$$NO + O_3 \rightarrow NO_2 + O_2 \qquad (1)$$

$$NO + O_3 \rightarrow NO_2{}^* + O_2 \qquad (2)$$

$$NO_2{}^* \rightarrow NO_2 + hv \qquad (3)$$

In the first reaction nitric oxide reacts with ozone and forms some portion of nitrogen dioxide and oxygen; in the second reaction NO reaction with ozone leads to generation of nitrogen dioxide in excited state. Deactivation of $NO_2{}^*$ leads to emission of photons. Amount of light emitted is proportional to amount of NO.

2   Materials

2.1   Plant Material and Cultivation

1. In this method plants such as tobacco, pea, and barley were used.
2. Hydroponic trays.
3. Air pumps.
4. Air diffusers.
5. Plant growth chamber with controlled conditions such as temperature, humidity, and light.
6. Hydroponic solution contains 1 mM NH_4NO_3, 250 µM $CaCl_2$, 100 µM FeEDTA, 1 mM $MgSO_4$, 100 µM H_3BO_3, 1.5 µM $CuSO_4$, 50 µM KCl, 10 µM $MnSO_4$, 0.1 µM Na_2MoO_4, 100 µM Na_2SiO_3, 2 µM $ZnSO_4$, and 1 mM KH_2PO_4 (see **Notes 1** and **2**).

2.2   Mitochondria Isolation

1. Mortar and pestle.
2. Sucrose.
3. Tetra-sodium pyrophosphate.
4. Polyvinylpyrrolidone.
5. Ethylenediaminetetraacetic acid KH_2PO_4.
6. Bovine serum albumin.
7. Ascorbic acid.
8. Kitchen blender.
9. Miracloth (for filtration).
10. $MgCl_2$.
11. HEPES.
12. Percoll density gradient.

2.3   Chemiluminescence Detection

1. Pure air and nitrogen gas.
2. Transparent cuvette.
3. Rubber tubes.

4. Charcoal column 1 m long, 3 cm internal diameter, particle size 2 mm.

5. Pure nitric oxide gas for calibration.

6. Flow controllers (FC-260, Tylan General, Eching, Germany).

7. Orbitor shaker.

8. Blue tag.

9. Razor blade.

10. Blotting paper.

11. Weighing balance.

2.4   Indirect Chemiluminescence Detection

1. 50 mM Vanadium (III) chloride in 1 M HCl.

2. 50 mM HEPES (pH 7.4).

3. 1–5 U of NOS, 1 mM L-arginine, 1 mM $MgCl_2$, 0.1 mM NADPH, and 12 μM BH_4.

2.5   Description of the Analyzer CLD 770 AL ppt

Figure 1 shows various components of the analyzer. Despite the fact that the CLD 770 AL ppt analyzer contains two reaction chambers in the small pre-chamber, NO reacts with ozone. The actual chemiluminescence reaction occurs in the main reaction

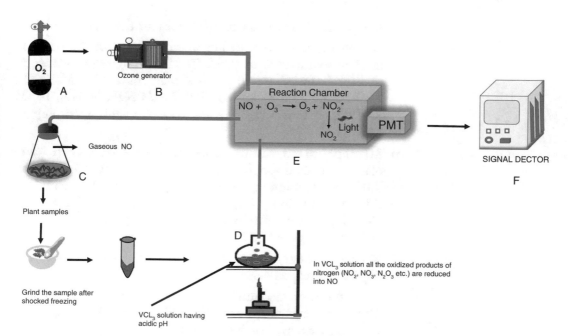

Fig. 1 Steps in direct and indirect chemiluminescence detection of nitric oxide. (**a**) Oxygen passes into ozone generator. (**b**) Ozone generator generates ozone and passes into reaction chamber. (**c**) Cells/root slices/leaf slices emit NO and pass into reaction chamber. (**d**) Injection of extracts into hot acidic VCLIII. (**e**) Reaction chamber where NO reacts with ozone and generation of NO chemiluminescence signal generation takes place. (**f**) Signal detector translates into ppb of NO

chamber. The reaction between NO and ozone takes place under low pressure. An external vacuum pump generates a reaction chamber with low pressure of 15 mbar. This low pressure is sufficient to drive both the gas sample and the ozone into the chamber. A photomultiplier collects the light and converts and amplifies current pulses. A microprocessor calculates the NO signal in the range of ppb. A custom-made software based on Visual Designer (PCI-20901SS, Ver. 4.0, Tuscon, Arizona, USA) was used to process the NO signal.

3   Methods

3.1   Plant Cultivation

1. Surface sterilize seeds with a mixture of Tween detergent (0.05 %) and H_2O_2 (0.015 %) and wash four times with distilled water. Water seeds and keep on wet filter paper for 48 h.

2. Grow *Nicotiana tabacum* cv. Gatersleben or nia plants in a vermiculite/sand mixture (2 parts vermiculite/1 part sand) for 2–3 weeks and from then transfer to hydroponic system for a further 3–4 weeks.

3. In case of barley and pea plants germinate seeds on filter paper after 2–3 days' growth transfer them directly to hydroponic system.

4. Use plants of growth stage 5–7 weeks for the experiments.

5. During the sand/vermiculite water plants with full-strength nutrient solution twice a week. The full-strength nutrient solution for "nitrate"-grown plants should contain 5 mM KNO_3, 1 mM $CaCl_2$, 1 mM $MgSO_4$, 0.025 mM NaFe-EDTA, 1 mM K_2HPO_4, 2 mM KH_2PO_4, and trace elements according to [26].

6. Adjust pH to 6.3 with KOH and transfer the vermiculite/sand plants into hydroponic tanks, each containing 15 plants in 8.0 l nutrient solution, and place in growth chambers with artificial illumination (HQI 400 W, Schreder, Winterbach, Germany) at a photon flux density of 300 µmol m^{-2} s^{-1} (PAR) and a day length of 16 h.

7. Maintain the day/night temperature regime of the chamber at 25/20 °C, respectively.

8. Flush the root medium with pre-moisturized air, using four 100 mm long micro-ceramic diffusers per tank (Europet Benelux, Gemert, The Netherlands).

3.2   Isolation of Mitochondria

1. All mitochondrial isolation steps should be carried out at 4 °C (*see* **Note 3**).

2. Roots and leaves need to be chopped with a sharp razor blade into approximately 0.5 cm slices and 2 g of slices/10 ml solu-

tion and placed in 50 ml plastic measuring cylinder and ground the tissue using Ultra Turrax.

3. Filtrate the homogenate using one layer of Miracloth and four layers of nylon mesh (80–100 μm).

4. Centrifuge the filtrate at $2000 \times g$ for 10 min. Discard the pellet.

5. Centrifuge the supernatant at $12,000 \times g$ for 30 min, and discard the supernatant.

6. The pellet should be dissolved by passing soft paintbrush/application of slight pressure with pipette on the pellet or by repeatedly rinsing the pellet with a small volume of medium using a Pasteur pipette. Finally suspend the pellet in 2 ml of suspension buffer.

7. Place the mitochondrial suspension on the discontinuous Percoll gradient. Required concentrations of gradients should be prepared by mixing specific concentration of Percoll in a Percoll buffer [27, 28]. More specifically—the first layer (from below) contains 3 ml of 60 % Percoll (v/v) and then overlay with 4 ml 45 % (v/v) and then overlay with 4 l of 28 % (v/v) Percoll and then on the top with 4 ml of 5 % (v/v) Percoll. Load Percoll gently with Pasteur pipette at 40° angle. Preparation of gradient solution can be done by using gradient mixture or by smooth pipetting.

8. The mitochondrial fraction appears at the interface between 45 and 28 % (v/v). Gently remove the layer with a Pasteur pipette, place it in a 50 ml centrifuge tube that contains 15 ml of suspension buffer, and centrifuge at $18000 \times g$ for 15 min. Discard the supernatant, resuspend the pellet in 15 ml suspension buffer, and centrifuge again at $18,000 \times g$ for 15 min.

9. Yellowish brown pellet containing the root mitochondria can be seen at the bottom of the tube.

3.3 Methods to Check Activity and Integrity of the Mitochondria

1. Monitoring mitochondrial activity: State 3/state 4 ratio is referred to as the respiratory control ratio (P:O). State 3 respiration means ADP-enhanced respiration. State 4 means respiration in the complete absence of the ATP synthesis. State 4 can be achieved by adding the ATP synthase inhibitor oligomycin (1 μg/ml). Oxygen uptake measurements for checking state 3/state 4 ratio can be done using oxygen electrode or by using the Microx TX2 oxygen-sensing device (PreSens Precision Sensing). A respiratory control ratio (P:O) of 3 means that mitochondria are well coupled; values lower than this indicate that they are only loosely coupled and that membranes are eventually damaged.

2. Peroxisomal contamination can be checked by adding 1 mM H_2O_2 to the mitochondrial suspension. Rates of oxygen evolution are proportional to the peroxisomal content.

3. Cytosolic contamination can be checked by measuring a cytosolic marker such as phosphoenolpyruvate carboxylase (PEPC) activity.

4. Thylakoid contamination can be checked by measuring the chlorophyll content in leaf mitochondria.

5. Western blots can be done by using antibodies against various marker enzymes of subcellular compartments, e.g., against peroxisomal protein KAT2 (3-ketoacyl-CoA thiolase-2) for checking peroxisomal contamination. Chloroplast contamination can be checked by antibody against large subunit of Rubisco [29].

3.4 Preparation of Root Segments

1. Wash roots with autoclaved deionized water.

2. On a clean glass plate spread root system.

3. Remove excess of water from root system by placing a blotting paper on the top of root system.

4. Cut lowest part of 2–3 cm lowest part of roots including root tips were cut into 3–5 mm segments, and place in a glass vessel containing 10 ml 20 mM HEPES-KOH, pH 7.0, 0.5 mM KNO_2, and 50 mM sucrose.

3.5 Preparation of Leaf Segments

Rinse freshly harvested leaves with deionized water and cut into 1 cm long and 1–3 mm wide segments excluding the midrib portion.

Vacuum infiltrate the leaf segments (1 g FW) for 2–3 min in 10 ml of 25 mM HEPES-KOH pH 7.4 and 0.5 mM $CaSO_4$.

Wash the segments and subsequently suspend in the same buffer for the NO measurements.

3.6 Measurement of Nitric Oxide

1. Place the vials containing root segments or leaf slices (1 g FW in 10 ml buffer) or mitochondrial suspensions (1.5–2.5 mg protein in 8 ml buffer) in a glass cuvette (1.0 l air volume) mounted on a rotary shaker (150 U min^{-1}).

2. Pull a constant flow of measuring gas (purified air or nitrogen) of 1.6 l min^{-1} through the cuvette, and through a cold moisture trap and subsequently through the chemiluminescence detector (CLD 770 AL ppt, Eco-Physics, Dürnten, Switzerland, detection limit 20 ppt; 20-s time resolution) by a vacuum pump connected to an ozone destroyer.

3. Make the measuring gas (air or nitrogen) NO free by conducting it through a custom-made charcoal column (1 m long, 3 cm internal diameter, particle size 2 mm).

4. Carry out calibration with NO-free air (0 ppt NO) and with various concentrations of NO (1–35 ppb) adjusted by mixing the calibration gas (500 ppb NO in nitrogen, Messer Griesheim, Darmstadt, Germany) with NO-free air. Use flow controllers (FC-260, Tylan General, Eching, Germany) to adjust all gas flows.

5. Maintain air temperature of the cuvette at 25 °C. Customize the cuvette lid to allow the injection of various solutions through a serum stopper directly into the sample without opening the cuvette or interrupting shaking cycles.

6. Record reading by using Visual designer-based software (Intelligent Instrumentation Inc., Tucson, USA).

3.7 Nitrate/Nitrite Trace Analysis ("Indirect chemiluminescence")

1. Inject 100–500 µl of the respective into a reducing reaction mixture [50 mM vanadium (III) chloride in 1 M HCl] at 90 °C, under continuous stirring.

2. Pass the emitted gas via 100 ml of 1 M KOH to protect the analyzer from HCl carried over (*see* **Note 4**).

3. Calculate the production of NO by subtracting the 0 time value which represents nonenzymatic NO production from suspension medium and cell components [30].

4. Measure NO from recombinant iNOS as a positive control. This iNOS assay must contain 50 mM HEPES (pH 7.4), 1–5 U of NOS, 1 mM L-arginine, 1 mM $MgCl_2$, 0.1 mM NADPH, and 12 µM BH_4.

4 Notes

1. While growing plants on hydroponics regularly check pH.

2. Change nutrient solution at least three times a week.

3. Always perform all steps of mitochondria isolation at 4 °C.

4. While doing indirect chemiluminescence always watch the setup. Overheating of vanadium (III) chloride can damage setup and instrument.

Acknowledgments

This work was supported by Ramalingaswami Fellowship funded to JGK by DBT. I thank Werner Kaiser, University of Wuerzburg, for introducing chemiluminescence method. RJ and PKP are currently funded by UGC Fellowship for doctoral studies.

References

1. Mur LAJ, Mandon J, Persijn S, Cristescu S, Moshkov I, Novikova G, Hall M, Hareen F, Hebelstrup K, Gupta KJ (2013) Nitric oxide in plants: an assessment of the current state of knowledge. AoB Plants 5:pls052

2. Gupta KJ, Fernie AR, Kaiser WM, Van Dongen JT (2011) On the origins of nitric oxide. Trends Plant Sci 16(3):160–168

3. Moreau M, Lindermayr C, Durner J, Klessig DF (2010) NO synthesis and signaling in plants–where do we stand? Physiol Plant 138(4):372–383

4. Wink DA, Mitchell JB (1998) Chemical biology of nitric oxide: insights into regulatory, cytotoxic, and cytoprotective mechanisms of nitric oxide. Free Radic Biol Med 25(4–5):434–456

5. Agurla S, Gayatri G, Raghavendra AS (2014) Nitric oxide as a secondary messenger during stomatal closure as a part of plant immunity response against pathogens. Nitric Oxide 43:83–96

6. Gupta KJ, Igamberdiev AU (2013) Recommendations of using at least two different methods for measuring NO. Front Plant Physiol 4:58

7. Mur LAJ, Mandon J, Cristescu SM, Harren FJ, Prats E (2011) Methods of nitric oxide detection in plants: a commentary. Plant Sci 181(5):509–519

8. Vitecek J, Reinohla V, Jones RL (2008) Measuring NO production by plant tissues and suspension cultured cells. Mol Plant 1: 270–284

9. Rockel P, Strube F, Rockel A, Wildt J, Kaiser WM (2002) Regulation of nitric oxide (NO) production by plant nitrate reductase *in vivo* and *in vitro*. J Exp Bot 53:103–110

10. Harper JE (1981) Evolution of nitrogen oxide(s) during *in vivo* nitrate reductase assay of soybean leaves. Plant Physiol 68:1488–1493

11. Klepper LA (1987) Nitric oxide emissions from soybean leaves during *in vivo* nitrate reductase assays. Plant Physiol 85:96–99

12. Gupta KJ, Stoimenova M, Kaiser WM (2005) In higher plants, only root mitochondria, but not leaf mitochondria reduce nitrite to NO, *in vitro* and *in situ*. J Exp Bot 56:2601–2609

13. Gupta KJ (2007) Nitric oxide in plants: Investigation of synthesis pathways and role in defense against avirulent *Pseudomonas*. Thesis, Würzburg, Germany

14. Chen J, Vandelle E, Bellin D, Delledonne M (2014) Detection and function of nitric oxide during the hypersensitive response in *Arabidopsis thaliana*: where there's a will there's a way. Nitric Oxide 43:81–88

15. Corpas FJ, Barroso JB, Carreras A, Valderrama R, Palma JM, Leon AM, Sandalio LM, del Rı´o AL (2006) Constitutive arginine-dependent nitric oxide synthase activity in different organs of pea seedlings during plant development. Planta 224:246–254

16. Planchet E, Sonoda M, Zeier J, Kaiser WM (2006) Nitric oxide (NO) as an intermediate in the cryptogein induced hypersensitive response a critical re-evaluation. Plant Cell Environ 29:59–69

17. Planchet E, Gupta KJ, Sonoda M, Kaiser WM (2005) Nitric oxide emission from tobacco leaves and cell suspensions: rate-limiting factors and evidence for the involvement of mitochondrial electron transport. Plant J 41:732–743

18. Gupta KJ, Kaiser WM (2010) Production and scavenging of nitric oxide by barley root mitochondria. Plant Cell Physiol 51(4):57

19. Conrath U, Amoroso G, Kohle H, Sultemeyer DF (2004) Non-invasive online detection of nitric oxide from plants and some other organisms by mass spectrometry. Plant J 38:1015–1022

20. Griveau S, Besson-Bard A, Bediqui F, Wendehenne D (2015) Electrochemical detection of nitric oxide in plant cell suspension. In: Walker J, Gupta KJ (eds) Methods in molecular biology. Springer, New York

21. Shibuki K (1990) An electrochemical microprobe for detecting nitric-oxide release in brain-tissue. Neurosci Res 9:69–76

22. Besson-Bard A, Griveau S, Bedioui F, Wendehenne D (2008) Real-time electrochemical detection of extracellular nitric oxide in tobacco cells exposed to cryptogein, an elicitor of defence responses. J Exp Bot 59:3407–3414

23. Weaver J, Porasuphatana S, Tsai P, Budzichowski T, Rosen GM (2005) Spin trapping nitric oxide from neuronal nitric oxide synthase: a look at several iron–dithiocarbamate complexes. Free Radic Res 39:1027–1033. doi:10.1080/10715760500231885

24. Murphy ME, Noack E (1994) Nitric oxide assay using hemoglobin method. Methods Enzymol 233:240–250. doi:10.1016/S0076-6879(94)33027-1

25. Cvetkovska M, Vanlerberghe GC (2012) Alternative oxidase modulates leaf mitochondrial concentrations of superoxide and nitric oxide. New Phytol 195

26. Johnson C, Stout P, Broyer T, Carlton A (1957) Comparative chlorine requirements of different plant species. Plant Soil 8:337–353

27. Vanlerberghe GC, Day DA, Wiskich JT, Vanlerberghe AE, McIntosh L (1995)

Alternative oxidase activity in tobacco leaf mito-chondria. Dependence on tricarboxylic acid cycle-mediated redox regulation and pyruvate activation. Plant Physiol 109:353–361

28. Nishimura M, Douce R, Akazawa T (1982) Isolation and characterization of metabolically competent mitochondria from spinach leaf protoplasts. Plant Physiol 669:916–920

29. Duncan O, Taylor NL, Carrie C, Eubel H, Kubiszewski-Jakubiak S, Zhang B, Narsai R, Millar AH, Whelan J (2011) Multiple lines of evidence localize signaling, morphology, and lipid biosynthesis machinery to the mitochon-drial outer membrane of Arabidopsis. Plant Physiol 157(3):1093–1113

30. Braman RS, Hendrix SA (1989) Nanogram nitrite and nitrate determination in environ-mental and biological materials by vanadium (III) reduction with chemiluminescence detec-tion. Anal Chem 61:2715–2718

31. Mur LAJ, Santosa IE, Laarhoven LJJ, Holton NJ, Harren FJM, Smith AR (2005) Laser pho-toacoustic detection allows *in planta* detection of nitric oxide in tobacco following challenge with avirulent and virulent *Pseudomonas syringae* pathovars. Plant Physiol 138:1247–1258

Chapter 3

Nitric Oxide Measurement from Purified Enzymes and Estimation of Scavenging Activity by Gas Phase Chemiluminescence Method

Aprajita Kumari, Alok Kumar Gupta, Sonal Mishra, Aakanksha Wany, and Kapuganti Jagadis Gupta

Abstract

In plants, nitrate reductase (NR) is a key enzyme that produces nitric oxide (NO) using nitrite as a substrate. Lower plants such as algae are shown to have nitric oxide synthase enzyme and higher plants contain NOS activity but enzyme responsible for NO production in higher plants is subjected to debate. In plant nitric oxide research, it is very important to measure NO very precisely in order to determine its functional role. A significant amount of NO is being scavenged by various cell components. The net NO production depends in production minus scavenging. Here, we describe methods to measure NO from purified NR and inducible nitric oxide synthase from mouse (iNOS), we also describe a method of measure NO scavenging by tobacco cell suspensions and mitochondria from roots.

Key words Nitrate reductase, Scavenging, Nitric oxide synthase, Mitochondria, Chemiluminescence

1 Introduction

The first preliminary evidence for NO emission was reported by Harper [1]. After few years of this finding, Dean and Harper [2] using GC and GC/MS found that nitrate reductase (NR) deficient soybean mutants are unable to generate NO. There findings suggest that NR is the key enzyme for NO production. Later Dean and Harper [3] found that NR requires NAD(P)H and pH optimum is 6.75. It has been shown that NR has capability to generate NO in vitro and in vivo [4]. It was found that NO is generated using MoCo (Molybdenum as cofactor) as cofactor [5]. This was evidenced by adding tungstate into the medium leads to reduced levels of NO from leaves [6]. NR can generate only 1 % of NO in respect to its activity [4]. The Km value for nitrite dependent NO production from NR is 100 μM [4]. It was found that Km value of nitrite dependent NO production from NR is also depends on

Kapuganti Jagadis Gupta (ed.), *Plant Nitric Oxide: Methods and Protocols*, Methods in Molecular Biology, vol. 1424,
DOI 10.1007/978-1-4939-3600-7_3, © Springer Science+Business Media New York 2016

intra cellular nitrate levels (Ki nitrate is 50 μM). It was found that antisense suppression of nitrite reductase NiR (nitrite reductase) leads to increased levels of nitrite and NO production [7]. Depending on species type NR exists in homodimer to tetramer (100–115 kDa). The catalytic reaction requires Heme, FAD, and MoCo. NR activity is controlled by phosphorylation and dephosphorylation (Figs. 1 and 2). For instance, transition from light to dark leads to phosphorylation of NR and leads to inhibition of its activity. Serine 543 is the site for phosphorylation of NR [8, 9] Ca^{2+} plays a role in phosphorylation [10]. NR can be inhibited by azide or cyanide [11, 12]. Addition of uncouplers leads increased NO activity [4].

Another enzyme that produces NO is nitric oxide synthase (NOS). In animal systems, there are three forms of NOS exist [13, 14]. These are neuronal NOS (nNOS), endothelial NOS (eNOS), and inducible NOS (iNOS), in addition to these there are also reports about presence mitochondrial NOS (mtNOS) [15] NOS is

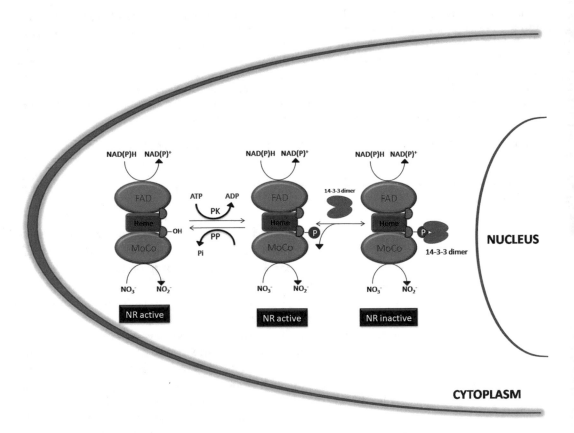

Fig. 1 An illustration of posttranslational modulation of nitrate reductase. NR structure contains three different functional and structural domains and those are connected by hinge regions. The Hinge-1 contains serine-543 phosphorylation site this connects the heme and the MoCo domain. The phosphorylated ser-motif is recognized by a 14-3-3 dimer leads to completely inactive complex, such condition leads to losing ability to transfer electrons from NAD(P)H

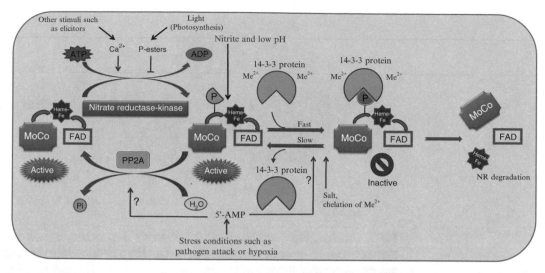

Fig. 2 A model for regulation of nitrate reductase by reversible binding of 14-3-3 and via phosphorylation and dephosphorylation. NR becomes active by phosphorylation and this process is further enhanced by the presence of nitrite which accumulates under hypoxia and low pH. NR degradation occurs when 14-3-3 binds for longer duration

a class of cytochrome P450 enzyme [14]. IT C-terminal oxygenase domain contains protoporphyrin IX domain heme iron and tetrahydrobiopterin BH_4 binding site for Arginine and oxygen (Fig. 3). Electrons are transferred from NAD(P)H to heme and to oxygen and then react with arginine guanidine group and producing *N*-hydroxy arginine (NOHA). This NOHA is further oxidized to NO and citrulline (Fig. 4).

In plants, numerous reports are available for NOS like activities for examples [16] has shown NOS activity in developing pea, it was found that this activity is located in peroxisomes [17]. Guo et al. [18] has identified NOS like protein in *Arabidopsis* and they named as AtNOS1. It is 60 kDa protein. AtNOS1 mutants have reduced levels of NO. Later it was shown to be involved in senescence and protection against oxidative damage [19, 20]. Later it was found out to be GTPase not a true NOS therefore this protein was names at AtNOA1 [20].

Chemiluminescence is a robust technique to measure NO emitted from various sources such as cell suspensions, leaves, purified NR [6] roots and mitochondria [21] and purified NOS [22]. By this technique, one can estimate scavenging activity very precisely. In this procedure, NO solution can be prepared by flushing NO gas into solution. Once this solution is ready, it can be injected into solution that is being evaluated for measurement of scavenging activity. The amount of NO that is recovered from total NO will be measured. Amount of NO added minus amount of NO liberated is equal to amount of NO scavenged.

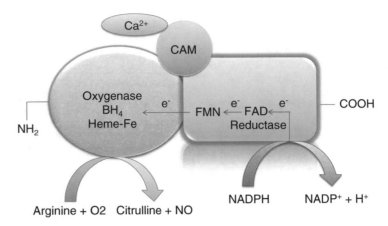

Fig. 3 Structural representation of NOS that contains reductase domain and Heme-Fe domain. These two domains are seperated by a calmodulin binding site and (figure modified from [20])

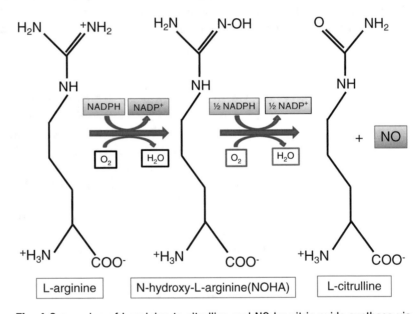

Fig. 4 Conversion of l-arginine to citrulline and NO by nitric oxide synthase via formation of N-hydroxy-l-arginine as (NOHA) as an intermediate

Here, we show procedure for measurement of NO from purified Maize NR and in inducible NOS from mice and estimation of scavenging by mitochondria of roots.

2   Materials

2.1   Plant Material and Cultivation

1. Barley seeds.

2. Hydroponic trays.

3. Air pumps.

4. Ciramic air diffusers.

5. Plant growth chamber with controlled conditions such as temperature, humidity, and light.

6. Hydroponic solution contain 1 mM NH_4NO_3, 250 μM $CaCl_2$, 100 μM FeEDTA, 1 mM $MgSO_4$, 100 μM H_3BO_3, 1.5 μM $CuSO_4$, 50 μM KCl, 10 μM $MnSO_4$, 0.1 μM Na_2MoO_4, 100 μM Na_2SiO_3, 2 μM $ZnSO_4$, and 1 mM KH_2PO_4.

2.2 Mitochondria Isolation

1. Mortar and pestle.

2. Sucrose.

3. Tetra-sodium pyrophosphate.

4. Polyvinylpyrrolidone.

5. Ethylenediaminetetraacetic acid KH_2PO_4.

6. *Bovine Serum Albumin.*

7. Ascorbic Acid.

8. Kitchen Blender.

9. Miracloth (for filtration).

10. $MgCl_2$.

11. HEPES.

12. Percoll density gradient.

2.3 Chemilumine scence Detection

1. Pure air and nitrogen gas.

2. Transparent cuvette.

3. Rubber tubes.

4. Charcoal column 1 m long, 3 cm internal diameter, particle size 2 mm.

5. Pure nitric oxide gas for calibration.

6. Flow controllers (FC-260, Tylan General, Eching, Germany).

7. Orbitor shaker.

8. Blue tag.

9. Razor blade.

10. Blotting paper.

11. Weighing balance.

12. NO gas 500 ppm.

2.4 Indirect Chemiluminescence Detection

1. 50 mM vanadium (III) chloride in 1 M HCl.

2. 50 mM HEPES (pH 7.4).

3. 1–5 U of NOS, 1 mM l-arginine, 1 mM $MgCl_2$, 0.1 mM NADPH, and 12 μM BH_4.

4. Glassware as shown in Chapter 2.

**2.5 Material
for Measurement
of NO
from iNOS and NR**

1. Recombinant iNOS (mouse) EC: 1.14.13.39 (Sigma N-2783).
2. 100 mM l-arginine.
3. 100 mM NADPH.
4. 200 mM Tetrahydrobiopterin (BH_4).
5. 500 mM L-NAME L-N^G-Nitroarginine methyl ester.
6. 100 mM HEPES.
7. 50 mM Vanadium III chloride.
8. Heat block.
9. Purified NR (Maize) (N-9523 Sigma).
10. 100 mM NADH.

3 Methods

**3.1 Measurement
of NO from NOS**

1. Pipette 5 ml of HEPES buffer in a small glass cuvette (10 ml volume).
2. Add 1 U of iNOS into the buffer.
3. Add 1 mM l-arginine. 1 mM $MgCl_2$, 12 μM BH_4.
4. Shake the container in a gas space cuvette 60 rpm/min until NO production saturates.
5. Inject 5 mM NOS inhibitor.
6. Wait until NO emission diminish.
7. Remove the cuvette immediately and freeze the buffer for nitrite analysis.
8. Inject 100 μl of sample into [50 mM vanadium (III) chloride in 1 M HCl] at 90 °C.
9. Pass the emitted gas through 100 ml of 1 M KOH to protect the analyzer from HCl carried over.
10. Subtract the 0 time value which represents nonenzymatic NO production from suspension medium and cell components.
11. Inject 1–10 μM nitrite into VCl III solution for estimation of nitrite.

**3.2 Measurement
of Scavenging Activity
of NO by Cell
Suspensions or
Mitochondria [21]**

1. Prepare define amount of NO by mixing 100 ml of distilled water with 95 ppm of NO under nitrogen atmosphere.
2. This will give 180 pmol/ml concentration of NO.
3. Take 2 ml of NO in a syringe and inject into 10 ml of buffer or that contain 1 g cell suspension or 1 g roots slices or 2 mg of mitochondria.
4. Stir the solution continuously in headspace cuvette that is continuously flushed with NO free air or nitrogen gas.

5. Measure the released NO for at least 20 min. Until peak reaches to 0.

6. Calculate the integral of the peak which reflects amount of that is released from actual NO added.

7. Added NO minus released NO gives total NO scavenged.

3.3 Measurement of NO from Purified NR

1. Take 3 ml of buffer in 5 ml of glass cuvette.

2. Add aliquot of pure NR, 1 mM nitrite.

3. Add 500 μM NADH into the cuvette and determine NO released by chemiluminescence method.

4. While doing measurement keep temperature of cuvette at 22 °C.

4 Notes

1. Aliquot purified enzymes, therefore freeze thawing will be avoided.

2. Temperature has great influence on NO emission from purified enzymes therefore keep the temperature constant while doing measurement.

3. Purified NO gas leakage can cause potential problems therefore always check pipes and fittings before and during the measurement.

4. Shaker speed should be constant throughout the measurement.

Acknowledgments

This work was supported by Ramalingaswami Fellowship funded to JGK by DBT. I thank Werner Kaiser, University of Wuerzburg for introducing chemiluminescence method.

References

1. Harper JE (1981) Evolution of nitrogen oxide(s) during in vivo nitrate reductase assay of soybean leaves. Plant Physiol 68:1488–1493

2. Dean JV, Harper JE (1986) Nitric oxide and nitrous oxide production by soybean and winged bean during the in vivo nitrate reductase assay. Plant Physiol 82:718–723

3. Dean JV, Harper JE (1988) The conversion of nitrite to nitrogen oxide(s) by the constitutive NAD(P)H- nitrate reductase enzyme from soybean. Plant Physiol 88:389–395

4. Rockel P, Strube F, Rockel A, Wildt J, Kaiser WM (2002) Regulation of nitric oxide (NO) production by plant nitrate reductase in vivo and in vitro. J Exp Bot 53:103–110

5. Harrison R (2002) Structure and function of xanthine oxidoreductase: where are we now? Free Radic Biol Med 33:774–797

6. Planchet E, Gupta KJ, Sonoda M, Kaiser WM (2005) Nitric oxide emission from tobacco leaves and cell suspensions: rate- limiting factors and evidence for the involvement of mito-

chondrial electron transport. Plant J 41: 732–743

7. Morot-Gaudry-Talarmain Y, Rockel P, Moureaux T, Quileré I, Leydecker MT, Kaiser WM, Morot-Gaudry JF (2002) Nitrite accumulation and nitric oxide emission in relation to cellular signalling in nitrite reductase antisense plants. Planta 215: 708–715

8. Bachmann M, Shiraishi N, Campbell WH, Yoo BC, Harmon AC, Huber SC (1996) Identification of Ser-543 as the major regulatory phosphoprotein site in spinach leaf nitrate reductase. Plant Cell 8:505–517

9. Hoff T, Truong HN, Caboche M (1994) Use of mutants and transgenic plants to study nitrate assimilation. Plant Cell Environ 17:489–506

10. Huber SC, Bachmann M, Huber JL (1996) Post-translational regulation of nitrate reductase activity: a role for Ca2+ and 14-3-3-proteins. Trends Plant Sci 12: 432–438

11. Notton BA, Hewitt EJ (1971) Reversible cyanide inhibition of spinach (*Spinaceaoleracea L.*) nitrate reductase and non-exchangeability *in vitro* of protein bound molybdenum and tungsten. FEBS Lett 18:19–22

12. Yamasaki H, Sakihama Y (2000) Simultaneous production of nitric oxide and peroxynitrite by plant nitrate reductase: *in vitro* evidence for the NR dependent formation. FEBS Lett 468:89–92

13. Stuehr DJ, Kwon NS, Nathan CF, Griffith OW, Feldman PL, Wiseman J (1991) N omega-hydroxy-L-arginine is an intermediate in the biosynthesis of nitric oxide from L-arginine. J Biol Chem 266(10):6259–6263

14. Alderton WK, Cooper CE, Knowles RG (2001) Nitric oxide synthases: structure, function and inhibition. Biochem J 357:593–615

15. Ghafourifar P, Asbury ML, Joshi SS, Kincaid ED (2005) Determination of mitochondrial nitric oxide synthase activity. Methods Enzymol 396:424–444

16. Corpas FJ, Barroso JB, Carreras A, Valderrama R, Palma JM, León AM, Sandalio LM, del Río LA (2006) Constitutive arginine-dependent nitric oxide synthase activity in different organs of pea seedlings during plant development. Planta 224:246–254

17. Barroso JB, Corpas FJ, Carreras A, Sandalio LM, Valderrama R, Palma JM, Lupianez JA, del Rio LA (1999) Localization of nitric-oxide synthase in plant peroxisomes. J Biol Chem 274:36729–36733

18. Guo FQ, Okamoto M, Crawford NM (2003) Identification of a plant nitric oxide synthase gene involved in hormonal signalling. Science 302:100–104

19. Guo FQ, Crawford NM (2005) Arabidopsis nitric oxide synthase1 is targeted to mitochondria and protects against oxidative damage and dark- induced senescence. Plant Cell 17:3436–3450

20. Crawford NM, Galli M, Tischner R, Heimer YM, Okamoto M, Mack A (2006) Response to Zemojtel et al: (2006) Plant nitric oxide synthase: back to square one. Trends Plant Sci 11(11):526–527

21. Gupta KJ, Stoimenova M, Kaiser WM (2005) In higher plants, only root mitochondria, but not leaf mitochondria reduce nitrite to NO, in vitro and in situ. J Exp Bot 56:2601–2609

22. Gupta KJ, Kaiser WM (2010) Production and scavenging of nitric oxide by barley root mitochondria. Plant Cell Physiol 51(4):576–584

Chapter 4

Localization of Nitric Oxide in Wheat Roots by DAF Fluorescence

Aakanksha Wany and Kapuganti Jagadis Gupta

Abstract

Nitric oxide is a free radical signal molecule. Various methods are available for measurement of NO. Out of all methods, fluorescent probes to localize NO is very widely used method. Diaminofluorescein in diacetate form (DAF-2DA) is most widely probe for NO measurement. This method is based on application of 4,5-diaminofluorescein diacetate (DAF-2DA) which is actively diffused into cells, once taken up by cells cytoplasmic esterases cleave the acetate groups to generate 4,5-diaminofluorescein; DAF-2. The generated DAF-2 can readily react with N_2O_3, which is an oxidation product of NO to generate the highly fluorescent DAF-2T (triazolofluorescein). There are various advantages and disadvantages associated with this method, but to its advantage in diffusion closely to NO producing sites, it is widely used for localization studies. Here, we describe method to make sections of the roots and localization of NO in roots subjected to hypoxic stress.

Key words Nitric oxide, Hypoxia, DAF fluorescence, Microtome, Microscope

1 Introduction

Nitric oxide is a reactive nitrogen species which play multiple roles in plants. There are various methods available for measurement of NO. Out of all the methods, Diaminofluorescence (DAF) method is widely used in research. In the year 2008, Kojma et al. designed and synthesized diaminofluoresceins (DAFs) as novel fluorescent indicators for NO. The fluorescent chemical transformation of DAFs is based on application of 4,5-diaminofluorescein diacetate (DAF-2DA) which is actively diffused into cells, once taken up by cells cytoplasmic esterases cleave the acetate groups to generate 4,5-diaminofluorescein; DAF-2. The generated DAF-2 can readily react with N_2O_3, which is an oxidation product of NO to generate the highly fluorescent DAF-2T (triazolofluorescein). In recent years, Diaminofluorescein (DAF) fluorescent dyes are most widely used fluorescent probes for detection of nitric oxide from various tissues of plants [1–5]. These dyes are commercially available from

Kapuganti Jagadis Gupta (ed.), *Plant Nitric Oxide: Methods and Protocols*, Methods in Molecular Biology, vol. 1424, DOI 10.1007/978-1-4939-3600-7_4, © Springer Science+Business Media New York 2016

various companies. The advantage of these dyes is it does not react with NO_2^-, NO_3^-, and peroxynitrite ($ONOO^-$). Despite of several disadvantages researchers widely use DAF due to following reasons. Using detection limit for NO by DAF-2 is low, at about 5 nm. Therefore, it is sensitive enough for low concentrations of NO produced by various cell types such as stomata and roots hairs. Having excitation of 495 nm is advantageous due to less damage to cells during exposure to this safer wavelength. DAF is not a toxic compound therefore physiological properties don not alter during the experiment. Like many methods DAF has several disadvantages. DAF-2 reacts with dehydroascorbic acid (DHA) and ascorbic acid (AA) and generate novel compounds those have similar profiles matching DAF2T [6]. This problem can be overcome the use of ascorbate oxidase, in this reaction ascorbate oxidase is reduced to DHA and water [7]. In another study by [8] found that DAF functions only under aerobic conditions and can not be used to measure NO from anoxic tissues. Another study has shown that Horseradish-peroxidase (HR-PO) plus H_2O_2 generates DAF fluorescence in vitro and also H_2O_2 plus apoplastic peroxidase (PO) also generates NO [9]. DAF does not react directly with NO but instead it reacts with various oxidized products of NO such as NO^+ and N_2O_3 [10–12]. It has been also suggested that DAF-2 reacts with ROS that leads to some intermediates which can react with NO leads to fluorescence [13]. Since plants produce ROS and NO in response to particular process or stress it is difficult to precisely detect NO using DAF [14–16]. It has been shown that phenolic compounds also scavenge fluorescence [17]. DAF is sensitive to light therefore proper handing is required during fluorescent measurements.

DAF-FM has been developed as more sensitive NO sensor to DAF-2DA (respectively, ~3 nM and ~5 nM [11]). It has also been suggested that the fluorescent signal of DAF-FM is not affected by pH above 5. However, when measuring the concentration of dissolved NO in water using DAF-FM, [18] noted that the fluorescence obtained with 300 nM NO was quenched with increasing pH so that the signals at pH 9 were around half those obtained at pH 5.5. The plant cell cytosol pH is usually around 7.5, with the apoplast and vacuole being in the region of pH 5.5, but intracellular pH can change dramatically during cellular processes such as the pathogen-elicited HR, root tip growth, nodulation, gibberellic acid, and abscisic acid signaling. Thus, potential pH changes should be considered when interpreting DAF-FM result.

Due to importance of localization capability, it is important to use DAF. Previously, it was shown that DAF fluorescence increase in response to pathogens [19], elicitors [20], symbiotic microbes [21], xylem differentiation [22], lignification [22], context with root development [23], and stomata function [24–25]. Here, we describe method to section the roots and localize NO (Fig. 1) in root tissues subjected to hypoxic stress.

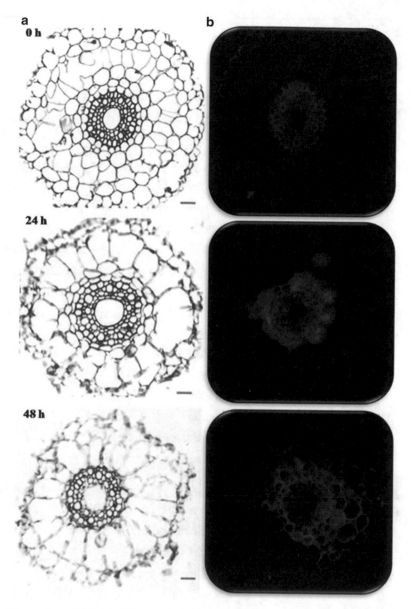

Fig. 1 Correlation between aerenchyma formation and DAF fluorescence compared by microtome sections made sections for DAF fluorescence. (**a**) Lysigenous aerenchyma in a transverse wax seminal root sections of wheat during hypoxia. Note the gas spaces in the cortex separated by bridges of cells spanning the spaces between the stele and epidermis; giving the spoked-wheel appearance in 24 and 48 h hypoxic treatment. Sections were stained with safranin and eosin. Scale bar = 250 μm. (**b**) DAF-2T fluorescence shown by seminal root hand sections of wheat during hypoxic treatment (24 and 48 h) observed under fluorescence microscope. Increased fluorescence can be seen in 24 and 48 h of treatments

2 Materials

2.1 Plant Material   Root segments (seminal roots from 21-days-old wheat seedlings).

2.2 Chemicals
Required

1. Sodium hypochlorite (4 %).

2. Hydroponics Medium (Hoagland's media): Macronutrients: 1 mM NH_4NO_3, 1 mM $CaCl_2$, 1 mM $MgSO_4$, 0.025 mM NaFe-EDTA, 1 mM K_2HPO_4, 2 mM KH_2PO_4; and micronutrients: 25 μM NaFe-EDTA, 15 μM H_3BO_3, 3 μM $MnCl_2 \cdot 4H_2O$, 0.25 μM $ZnSO_4 \cdot 7H_2O$, 0.1 μM $CuSO_4 \cdot 5H_2O$, and 0.04 μM Na_2MO_4.

3. 100 mM HEPES buffer (pH 7.2).

4. 10 % Glycerol.

5. DAF-FM dye (10 μM–working concentration).

6. Carboxy-PTIO potassium salt (200 μM–working concentration).

7. Cold FAA (formaldehyde–acetic acid–ethanol) fixative.

8. Ethanol dehydration series (10, 20, 30, 40, 50, 60, 70, 80, and 90 %).

9. Eosin stain (0.1 %).

10. Tert-butanol series (TBA-10, 20, 35, 55, 75, 85, 90, and 100 %).

11. Paraplast wax chips.

12. Xylene.

13. Safranin stain (0.1 %).

14. DPX mountant.

2.3 Equipments
Required

1. Phytotron (Plant growth chambers; 16 h photoperiod, day/night temperatures of 23–18 °C and relative humidity 60 %).

2. Anoxic chambers.

3. Aeration pumps.

4. Aeration pipes with needles.

5. Fluorescence microscope (Nikon ECLIPSE-80i, Japan).

6. Incubator oven (55 °C).

7. Slide warming table (70 °C).

8. Floatation bath (60 °C).

9. Hot plate (40 °C; overnight).

10. Rotary microtome (RM2265, Leica Biosystems, Germany).

11. Vacuum Pumps and Desiccators.

2.4 General Items
Required

1. Nitrogen gas cylinders.

2. Blotting paper sheets.

3. Petri dishes.

4. Hydroponic plastic trays.

5. Razor blade.

6. Forceps.

7. One-end frosted slides.

8. Cover glass.

9. Brush (0 size).

10. Plastic moulds.

11. Embedding rings.

12. Glass marker pens/Stickers/Tags/Sticking wide tapes.

13. Milli-Q water and sterile Milli-Q water.

3 Methodology/Protocol

3.1 Surface Sterilization of Seeds

1. Surface sterilize the seeds with 4 % sodium hypochlorite for 10 min and wash four to five times with distilled water by shaking the seeds until the smell of hypochlorite stops.

2. Prepare the petri dishes by keeping moist blotting paper sheets (circle shaped) for seed germination.

3. Place the surface sterilized seeds into the petri dishes and cover it with two layers of moistened blotting paper sheets.

4. Wrap up the petri dishes containing seeds with aluminum foil to maintain the dark conditions and incubate at 23 °C in growth chamber for 3 days.

3.2 Preparation of Hoagland's Media and Plant Growth

1. Prepare 1/4, 1/2, and full strength nutrient solution containing macronutrients and micronutrients in Milli-Q water and set the pH at 6.3.

2. Transfer 3-days-old germinated seedlings into the hydroponics trays containing 1/4, 1/2, and full strength nutrient solutions gradually allowing seedlings to adapt.

3. Provide first ¼ and later ½ strength nutrient solutions within 2 days interval after transferring the germinated seedlings to the hydroponics tray.

4. Provide aeration to the hydroponics medium using aeration pumps and keep the hydroponics trays with the germinated seedlings into the phytotron set at controlled conditions.

5. Consecutively change the nutrient solution to full strength after every 3–4 days interval till 21 days growth.

3.3 Hypoxia/Anoxia Treatment

1. Transfer the 21-days-old seedlings to hypoxic chambers for 24 and 48 h with their respective nutrient solutions.

2. Flash flush nitrogen gas three times a day for 1 h and ensure tightness to the chamber to keep hypoxic condition.

3. After hypoxic treatment, rinse the whole roots with deionized water and blot dry.

3.4 In Vivo NO Estimation

1. For sampling, excise root segments (2 cm long) from seminal roots of 21-days-old wheat seedlings both from control (normoxic) and treated (hypoxic; 24 and 48 h) samples.

2. Keep the freshly cut root segments in petri dishes containing nutrient solution.

3. Prepare cross sections from root segments (10 mm) by hand sectioning with a sharp razor blade (*see* **Note 4**).

4. Transfer these fine root sections to a 1.5 ml eppendorf tube (amber) that contains 100 mM HEPES buffer (pH 7.2), add 10 μM DAF-FM dye and incubate for 10 min in dark.

5. Wash the sections three to four times using 100 mM HEPES buffer and transfer onto the slide and mount with a drop of 10 % glycerol with a cover glass.

6. In another tube, add 10 μM DAF-FM and 200 μM cPTIO to the sections under same set of conditions, wash and prepare slides in the same way. This will serve as a control to find the observed fluorescence is due to NO or not (*see* **Notes 1–3**).

7. Observe the sections in a fluorescence microscope with 495 nm excitation and 515 nm emission wavelengths (Fig. 1).

3.5 Aerenchyma Studies/Microtomy Sectioning to Correlate NO Production and Aerenchyma Formation

We modified protocols from previous studies [26, 27] on microtomy sectioning and microscopy (Fig. 2).

1. Cut the root segments from the whole roots into 1 cm size and immediately immerse in cold fixative FAA (formaldehyde–acetic acid–ethanol) for 24 h at 4 °C.

2. Subject the vials that containing fixed root tissues to moderate vacuum for 15 min to remove the air out of the tissues.

3. Again, replace the vials with fresh fixative after vacuuming.

4. Start the dehydration series by treating the fixed tissues with the increasing ethanol series (10–90 %) to dehydrate the tissue, incubate at each step for an hour, and replace the old solution after incubation with the next solution in the series using a plastic transfer pipette.

5. At the last step of 90 % ethanol, add a drop of 0.1 % eosin stain and incubate for 30 min in order to distinguish the root tissues and replace with increasing *tert*-butanol (TBA) (10, 20, 35, 55, 75, 85, 90, and 100 %) series, incubate at each step for 30 min. Do the TBA series from 75–100 % in the incubation oven set at 50–60 °C as pure TBA crystallizes below 25 °C.

6. When the tissues are in 100 % TBA, incubate the vials for 2 h. Then, add Paraplast wax chips to the half-filled vials with TBA in the incubator oven set at 55 °C. Leave the samples in the oven overnight, so that all the TBA gets evaporated.

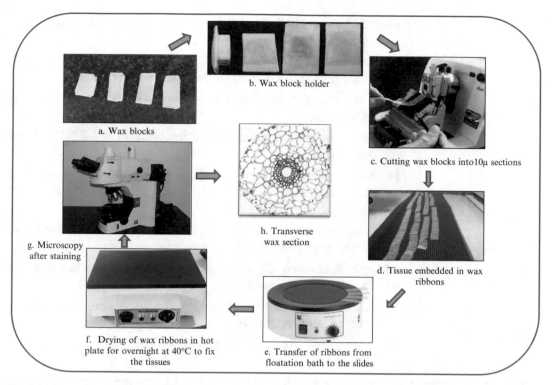

Fig. 2 (**a**) Preparing wax blocks; (**b**) filling of wax into the wax block holder; (**c**) setting the microtome at 10 μm, fixing the wax blocks holder into the microtome and cutting the tissue sections; (**d**) wax ribbons containing embedded tissues; (**e**) ribbons were gently moved in to the slide from the floatation bath; (**f**) drying of slides overnight over hot plate; (**g**) observing under the microscope after staining; (**h**) and final view of the transverse section (T.S. of wheat seminal root shown here)

7. The next day, pour away the TBA/wax mixture (if left) in a waste container. Now, replace the old wax with fresh molten wax every 10–12 h for 3 days, so as to ensure maximum penetration into the tissues to allow trouble free sectioning. This step completes the embedding of tissues in wax.

8. The next step is to cast the wax blocks. Use moulds for casting wax blocks. Keep the empty moulds over the slide warming table in the turned off condition. Add a small amount of molten wax to that mould so that it solidifies and then transfer the wax vials containing tissues to half-filled moulds over the warming table in the turned on condition.

9. With the help of warm forceps, place the tissues in the correct position in the mould and turn off the warming table.

10. Then, leave the mould in this position, and let it solidify. Keep the moulds in ice and store at 4 °C.

11. Cut the wax moulds in the specific positions where the root tissues were placed.

12. Fix the wax blocks containing root tissues into the embedding rings with hot wax.

13. Set the rotary microtome at 10 μm, place the embedding ring in the proper place and position and start sectioning.

14. The wax ribbons will come out continuously from the microtome.

15. Put those ribbons on the floatation bath set at 50 °C with the help of a brush.

16. Slowly take those ribbons by dipping the one-end frosted slides into the bath on the slide and heat fix them in the hot plate overnight at 40 °C.

17. Also, for ensuring correct section cutting, observe a small section of wax ribbon in the microscope.

18. After heat fixing, dip the slides containing ribbons in xylene for 15 min and let it air dry.

19. Counterstain the dried slides with 0.1 % safranin stain for 30 s, wash off and let it air dry.

20. Permanently fix the slides with a drop of DPX mountant.

4 Notes

1. DAF is light sensitive therefore always cover the DAF vial with aluminum foil or store in dark vial.

2. Switch o UV lamp of microscope 10 min before experiment.

3. Always wear gloves while doing DAF measurement.

4. DAF sectioning needs a fresh blade, always be alert while making sections.

References

1. Prats E, Mur LAJ, Sanderson R, Carver TLW (2005) Nitric oxide contributes both to papilla-based resistance and the hypersensitive response in barley attacked by *Blumeria graminis* f. sp. *Hordei*. Mol Plant Pathol 6:65–78

2. Prats E, Carver TLW, Mur LAJ (2008) Pathogen-derived nitric oxide influences formation of the appressorium infection structure in the phytopathogenic fungus *Blumeria gramini*. Res Microbiol 159:476–480

3. Foissner I, Wendehenne D, Langebartels C, Durner J (2000) In vivo imaging of an elicitor-induced nitric oxide burst in tobacco. Plant J 23:817–824

4. Krause M, Durner J (2004) Harpin inactivates mitochondria in Arabidopsis suspension cells. Mol Plant Microbe Interact 17:131–139

5. Lamotte O, Gould K, Lecourieux D, Sequeira-Legrand A, Lebrun-Garcia A, Durner J, Pugin A, Wendehenne D (2004) Analysis of nitric oxide signaling functions in tobacco cells challenged by the elicitor cryptogein. Plant Physiol 135:516–529

6. Zhang X, Kim WS, Hatcher N, Potgieter K, Moroz LL, Gillette R, Sweedler JV (2002) Interfering with nitric oxide measurements. 4,5-Diaminofluorescein reacts with dehydroascorbic acid and ascorbic acid. J Biol Chem 277:48472–48478

7. Kim WS, Ye XY, Rubakhin SS, Sweedler JV (2006) Measuring nitric oxide in single neurons by capillary electrophoresis with laser-induced fluorescence: use of ascorbate oxidase in diaminofluorescein measurements. Anal Chem 78:1859–1865

8. Planchet E, Kaiser WM (2006) Nitric oxide (NO) detection by DAF fluorescence and chemiluminescence: a comparison using abiotic and biotic NO sources. J Exp Bot 57: 3043–3055

9. Rümer S, Krischke M, Fekete A, Mueller MJ, Kaiser WM (2012) DAF-fluorescence without NO: elicitor treated tobacco cells produce fluorescing DAF-derivatives not related to DAF-2 triazol. Nitric Oxide 27(2):123–135

10. Kojima H, Nakatsubo N, Kikuchi K, Kawahara S, Kirino Y, Nagoshi H, Hirata Y, Nagano T (1998) Detection and imaging of nitric oxide with novel fluorescent indicators: diaminofluoresceins. Anal Chem 70:2446–2453

11. Kojima H, Sakurai K, Kikuchi K, Kawahara S, Kirino Y, Nagoshi H, Hirata Y, Nagano T (1998) Development of a fluorescent indicator for nitric oxide based on the fluorescein chromophore. Chem Pharm Bull 46:373–375

12. Espey MG, Miranda KM, Thomas DD, Wink DA (2001) Distinction between nitrosating mechanisms within human cells and aqueous solution. J Biol Chem 276:30085–30091

13. Jourd'Heuil D (2002) Increased nitric-oxide dependent nitrosylation of 4,5-diamino fluorescein by oxidants: implications for the measurement of intracellular nitric oxide. Free Radic Biol Med 33:676–684

14. Delledonne M, Zeier J, Marocco C, Lamb C (2001) Signal interactions between nitric oxide and reactive oxygen intermediates in the plant hypersensitive disease resistance response. Proc Natl Acad Sci U S A 98:13454–13459

15. Desikan R, Cheung MK, Bright J, Henson D, Hancock JT, Neill SJ (2004) ABA, hydrogen peroxide and nitric oxide signalling in stomatal guards. J Exp Bot 55:205–212

16. Gupta KJ, Mur LAJ, Brotman Y *Trichoderma asperelloides* suppresses nitric oxide generation elicited by *Fusarium oxysporum* in *Arabidopsis* roots. Mol Plant Microbe Int 27(4):307–314

17. Uhlenhut K, Hogger P (2012) Pitfalls and limitations in using 4,5-diaminofluorescein for evaluating the influence of polyphenols on nitric oxide release from endothelial cells. Free Radic Biol Med 52:2266–2275

18. Vitecek J, Reinohl V, Jones RL (2008) Measuring NO production by plant tissues and suspension cultured cells. Mol Plant 1:270–284

19. Delledonne M, Xia Y, Dixon RA, Lamb C (1998) Nitric oxide functions as a signal in plant disease resistance. Nature 394:585–588

20. Zeidler D, Zahringer U, Gerber I, Dubery I, Hartung T, Bors W, Hutzler P, Durner J (2004) Innate immunity in Arabidopsis thaliana: lipopolysaccharides activate nitric oxide synthase (NOS) and induce defense genes. Proc Natl Acad Sci U S A 101(44): 15811–15816

21. Baudouin E, Pieuchot L, Engler G, Pauly N, Puppo A (2006) Nitric oxide is formed in Medicago truncatula-Sinorhizobium meliloti functional nodules. Mol Plant Microbe Interact 19(9):970–975

22. Gabaldon C, Gomez Ros LV, Pedreno MA, Ros Barcelo A (2005) Nitric oxide production by the differentiating xylem of Zinnia elegans. New Phytol 165(1):121–130

23. Begara-Morales JC, Chaki M, Sanchez-Calvo B, Mata-Pérez C, Leterrier M, Palma JM, Barroso JB, Corpas FJ (2013) Protein tyrosine nitration in pea roots during development and senescence. J Exp Bot 64(4):1121–1134. doi:10.1093/jxb/ert006, Epub 2013 Jan 28

24. Neill SJ, Desikan R, Clarke A, Hancock JT (2002) Nitric oxide is a novel component of abscisic acid signaling in stomatal guard cells. Plant Physiol 128:113–116

25. Agurla S, Gayatri G, Raghavendra AS (2014) Nitric oxide as a secondary messenger during stomatal closure as a part of plant immunity response against pathogens. Nitric Oxide 43:89–96

26. Ruzin SE (1999) Plant microtechniques and microscopy. Oxford University Press, New York

27. Peterson RL, Peterson CA, Melville LH (2008) Teaching plant anatomy. NRC Press, Ottawa, Canada. ISBN 978-0-660-19798-2

Chapter 5

Nitric Oxide (NO) Measurements in Stomatal Guard Cells

Srinivas Agurla, Gunja Gayatri, and Agepati S. Raghavendra

Abstract

The quantitative measurement of nitric oxide (NO) in plant cells acquired great importance, in view of the multifaceted function and involvement of NO as a signal in various plant processes. Monitoring of NO in guard cells is quite simple because of the large size of guard cells and ease of observing the detached epidermis under microscope. Stomatal guard cells therefore provide an excellent model system to study the components of signal transduction. The levels and functions of NO in relation to stomatal closure can be monitored, with the help of an inverted fluorescence or confocal microscope. We can measure the NO in guard cells by using flouroprobes like 4,5-diamino fluorescein diacetate (DAF-2DA). This fluorescent dye, DAF-2DA, is cell permeable and after entry into the cell, the diacetate group is removed by the cellular esterases. The resulting DAF-2 form is membrane impermeable and reacts with NO to generate the highly fluorescent triazole (DAF-2T), with excitation and emission wavelengths of 488 and 530 nm, respectively. If time-course measurements are needed, the epidermis can be adhered to a cover-glass or glass slide and left in a small petri dishes. Fluorescence can then be monitored at required time intervals; with a precaution that excitation is done minimally, only when a fluorescent image is acquired. The present method description is for the epidermis of *Arabidopsis thaliana* and *Pisum sativum* and should work with most of the other dicotyledonous plants.

Key words 4,5-Diamino fluorescein diacetate, Nitric oxide, Reactive nitrogen species, Stomatal guard cells, Time-course measurement

1   Introduction

Nitric oxide is a free gas and can move cross biological membranes to exert its functions at subcellular compartments [1]. Initially, NO was discovered as a signaling component in animal systems during the relaxation of vascular smooth muscle, neurotransmission, immunological, and inflammatory responses. In plants, NO can regulate many developmental and physiological processes such as seed dormancy germination, root development, leaf senescence, and stomatal movements [2]. NO and its derived molecules collectively called as reactive nitrogen species (RNS) are involved in adaptive responses to abiotic and biotic stresses like drought, high salinity, chilling,

Kapuganti Jagadis Gupta (ed.), *Plant Nitric Oxide: Methods and Protocols*, Methods in Molecular Biology, vol. 1424, DOI 10.1007/978-1-4939-3600-7_5, © Springer Science+Business Media New York 2016

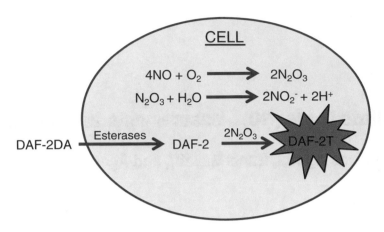

Fig. 1 Illustration of the reaction mechanisms leading to NO-dependent DAF-2T formation in plant cells

heat, and pathogen attack [3, 4]. At below 1 μmol/l, NO exhibits a half-life of minutes to hours. Such a long half-life makes NO accessible to deeper tissue layers. In contrast, at higher concentrations, NO has a short half-life, in the order of seconds [1].

Usage of cell permeable fluorescent dyes such as DAF-FM and DAF-2DA for the detection of NO has been tremendously changed the field of NO biology. Kojima et al. developed 4,5-diaminofluorescein (DAF-2) as a novel fluorescent indicator for NO [5, 6]. On application of 4,5-diaminofluorescein diacetate (DAF-2DA) is readily taken up into cells, where cytoplasmic esterases remove the acetate groups to generate 4,5-diamino fluorescein (DAF-2) preventing movement back out of the cell. DAF-2 can react with N_2O_3 (an oxidation product of NO) to generate the highly fluorescent DAF-2T (triazolofluorescein) ([7, 8] and Fig. 1). Besides direct monitoring of NO in the target cells, the role of NO is often demonstrated by the use of NO modulation by using scavengers like (2-phenyl-4,4,5,5-tetramethyl imidazoline-1-oxyl 3-oxide, cPTIO) and donors like sodium nitroprusside (SNP) and S-nitrosoglutathione (GSNO). Further, inhibitors of nitrate reductase (NR) such as sodium tungstate and nitric oxide synthase (NOS) N-nitro-l-Arg-methyl ester (L-NAME) are also used to examine the production and importance of NO in plant cells.

It is well known that NO is one of the important secondary messengers along with reactive oxygen species, free calcium and pH, during stomatal closure by several abiotic as well as biotic factors [9–11]. Monitoring of NO in guard cells of detached epidermis is routinely done in several laboratories, because of the simplicity of the system. The epidermis is single layered and the guard cells stand out because of their large size. As a result, stomatal guard cells have become quite popular model systems, for studying signaling components of plant cells. We have demonstrated that a rise in NO of guard cells is common during stomatal

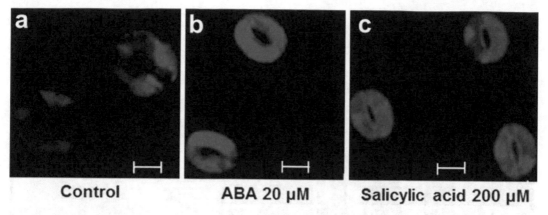

Fig. 2 Confocal images of DAF-2DA fluorescence in *Pisum sativum* guard cells. When treated with 20 μM of abscisic acid (ABA) and 200 μM salicylic acid, the fluorescence increased after 20 min (*bar* represents 20 μm). The panels are control (**a**), ABA (**b**) and salicylic acid (**c**)

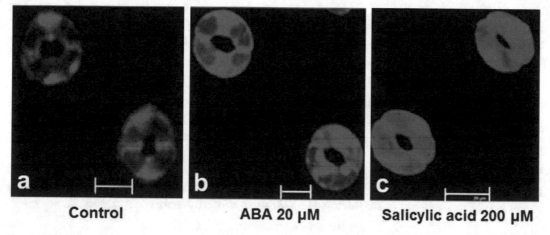

Fig. 3 DAF-2DA fluorescence in *Arabidopsis thaliana* guard cells, after treatment with 20 μM of abscisic acid (ABA) and 200 μM salicylic acid (*bar* represents 20 μm). The panels are control (**a**), ABA (**b**) and salicylic acid (**c**)

closure by not only abscisic acid (ABA) (Figs. 2, 3, and 4), but also even bicarbonate or elicitors like chitosan [12–14]. The role of NO in guard cells during stomatal closure is corroborated by using its scavengers of NO, promoters/inhibitors of NO production.

2 Materials

2.1 Buffers

1. Use 10 mM MES–KCl buffer pH-7.0 for opening of the stomata in *Pisum sativum* and pH 6.15 for *Arabidopsis thaliana* (*see* **Note 1**).

2. Prepare 100 mM MES [2-(*N*-morpholinoethanesulfonic acid] with ultrapure water and adjust the pH with potassium hydroxide (KOH) to the desired value.

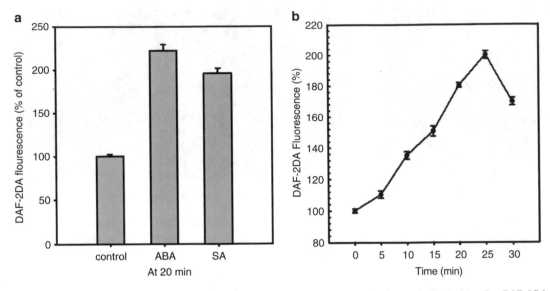

Fig. 4 (**a**) The increase in NO fluorescence in *Arabidopsis thaliana* guard cells, as indicated by the DAF-2DA fluorescence. The epidermal strips are treated with 20 μM of abscisic acid (ABA) or 200 μM salicylic acid (SA) for 20 min. (**b**) Increase in fluorescence with time in *Pisum sativum* guard cells, after treatment with 20 μM of abscisic acid up to 30 min

3. Take 25 ml of 100 mM MES–KOH solution and 12.5 ml of 1 M KCl in a volumetric flask and make up to 250 ml with ultrapure water to make 10 mM MES–KOH pH 7.0, and 50 mM KCl.

2.2 Fluorescent Probe

1. DAF-2DA is a NO specific fluorescent dye, available from InVitrogen or Calbiochem. Prepare a 5 mM stock solution by dissolving DAF-2DA in dimethyl sulfoxide (DMSO).

2. Store the stock solutions at −20 °C.

2.3 Light Source for Stomatal Opening

1. Use a bank of tungsten lamps as a light source for opening stomata. Filter the light through water jacket (*see* **Note 2**). Measure the photon flux with a Li-Cor quantum sensor (Li-Cor Instruments Ltd, Lincoln, NE, USA).

2. The room temperature should be maintained at 25 ± 1 °C, so as to optimize the function of stomatal guard cells.

2.4 Plant Growth

1. The pea plants are grown in a greenhouse (average day/night temperature of about 30/20 °C and photoperiod of 12 h) and are watered twice daily. The second to fourth completely unfolded leaves were collected from 2- to 3-week-old plants for epidermal bioassays.

2. Wild-type and various mutants of the Columbia ecotype of *Arabidopsis thaliana* were sown in soilrite and grown (16 h photoperiod, 100 μE/m²/s, 22 °C, 60 % relative humidity) in controlled environment growth chambers.

2.5 Microscope

1. Use either a standard microscope glass slide thick (0.13–0.16 mm) cover glass slip to adhere the epidermal strips. Watch the stomatal epidermis under a microscope. If necessary (e.g., in case of Arabidopsis) an adhesive like Telesis V (Premiere Products Inc., Pacaima, California, USA) may be used to adhere the abaxial epidermis to the glass slide.

2. Bright field microscope with 10X objective lens is sufficient to watch the stomatal movements but a microscope with attached camera is preferred. Captured and saved images of epidermis can be used for measurement of stomatal aperture, whenever convenient.

3. Either up-right fluorescent microscope or inverted fluorescent microscope (e.g., Optiphot-2, Nikon, Tokyo, Japan), with a monochrome high-resolution digital cooled CD camera (Cool-SNAP cf, Photometrics, Roper Scientific) or high resolution confocal laser scanning microscope (CLSM) to capture the fluorescent images of DAF-2DA fluorescence (filter: excitation 465–495 nm, emission 515–555 nm).

4. Capturing images has to be done in dark room as light may cause excitation of dye and it may give false results.

5. NIH Image or Scion image software is used for Windows to quantify the relative fluorescence emission of guard cells.

3 Methods

3.1 Bioassays of Stomatal Opening in Epidermal Strips

1. Peel off the abaxial (lower) epidermis of pea (*Pisum sativum*) or *Arabidopsis thaliana* leaves (*see* **Note 3**) and cut into strips of ca. 0.16 cm².

2. Transfer the epidermal strips to 3 cm diameter petri dishes or 6-well plate containing 3 ml of 10 mM MES–KOH pH 7.0 and 50 mM KCl buffer.

3. Expose epidermal strips for 3 h to the light, provided by light source. In *Arabidopsis thaliana*, the leaves are small so we have to incubate whole intact leaves under light with their lower epidermis exposed directly to the light (*see* **Note 4**).

4. Watch under bright field microscope for stomatal opening. After ascertaining sufficient stomatal opening (the average opening in pea plants is about 6.5–8.5 and in Arabidopsis wild type, it is about 3.5–4.5 μm), one can go for NO fluorescent measurement studies.

3.2 Monitoring NO in the Guard Cells of Pisum sativum and Arabidopsis thaliana

1. Load the epidermal strips with 20 μM DAF-2DA for 30 min (*see* **Note 5**) in opening buffer containing 0.05 % Pluronic F-127 (*see* **Note 6**), in dark at 25 ± 1 °C.

2. Rinse the strips quickly with three changes of incubation buffer to wash off the excessive fluorophore (*see* **Note 7**). Treat the strips with test compounds.

3. Monitor the epidermal strips under fluorescence microscope (as described in Subheading 2.5) to observe the fluorescence of DAF-2DA (*see* **Note 8**) and capture the images in JPEG or TIFF file format for further analysis [12–14].

3.3 Image Acquisition and Analysis

1. The levels of the fluorescence in the images can be quantified by using NIH Image or Scion image software for windows.

2. Import images to the NIH software (http://rsb.info.nih.gov/ij/index.html) and open as TIFF files, draw a square or oval box on the image window using the cursor and calculate the intensity of fluorescence by analyzing the pixels of the square box in the fluorescent image.

3. The mean values of square area box obtained by taking the pixels within the given fluorescence image window. Take "n" different pixel intensities of the square box of the same size in the nonfluorescent area, which can be used as the control (background).

4. The pixel intensity value of fluorescent guard cells was recorded as (X) and the background of the fluorescence images as (Y). The difference of the background and area of interest was calculated and Y–X gives the actual intensity of the fluorescent image.

5. The intensity of pixels in the control/beginning of the experiment is taken as 100 %. Based on the % of control, we can do analysis with various treatments.

3.4 Statistical Analysis

It is essential to repeat each experiment three to five times to ensure reproducibility. Make a statistical analysis of data by using SigmaPlot, which gives mean and error values. These values can also be used to draw histograms.

4 Notes

1. The MES–KCl buffer pH strength and composition may vary according to the plant species. It is advisable to standardize the best pH and composition suitable for each plant species.

2. An efficient water jacket is essential to filter the light, before it reaches the epidermis and the water in the jacket should not get heated, as temperature would affect stomatal movements.

3. In case of *Pisum sativum* as a plant material, the abaxial (lower) epidermis can be peeled off with a fine forceps. The epidermis need to be cut into strips of ca. $0.16\ cm^2$. In case of *Arabidopsis thaliana*, the leaves are very small and it is difficult to remove abaxial epidermis with forceps. So prepare paradermal sections of abaxial epidermis of Arabidopsis by mounting the epidermal sections on glass cover slips with the help of medical adhesive, Telesis V, and sections were allowed to open under light.

4. The light intensity to open the stomata also vary depend on the species. We generally use 250 and 150 $\mu mol/m^2/s$ light intensity for pea and Arabidopsis, respectively.

5. Over-time incubation of epidermis in the fluorophore results in oversaturation.

6. Pluronic F-127 is a detergent, which helps to permeate the plasma membrane and allow the dye into the cell.

7. Wash the excessive fluorophore two to three times with fresh buffer, otherwise the back ground will be high.

8. Keep the laser or Hg lamp shutter closed, except during the capture of images. In time course experiments with required time intervals, capturing images on time is important to get the best results.

Acknowledgments

Our work on guard cells is supported by grants to ASR of a J C Bose National Fellowship (No. SR/S2/JCB-06/2006) from the Department of Science and Technology and another from Department of Biotechnology, both in New Delhi. We also thank DBT-CREBB, DST-FIST, and UGC-SAP-CAS for support of infrastructure in Department/School.

References

1. Procházkova D, Wilhelmová N (2011) Nitric oxide, reactive nitrogen species and associated enzymes during plant senescence. Nitic oxide 24:61–65

2. Gayatri G, Agurla S, Raghavendra AS (2013) Nitric oxide in guard cells as an important secondary messenger during stomatal closure. Front Plant Sci 4:1–11

3. Delledonne M, Xia Y, Dixon RA et al (1998) Nitric oxide functions as a signal in plant disease resistance. Nature 394:585–588

4. Agurla S, Gayatri G, Raghavendra AS (2014) Nitric oxide as a secondary messenger during stomatal closure as a part of plant immunity response against pathogens. Nitric Oxide 43:89–96

5. Kojima H, Nakatsubo N, Kikuchi K et al (1998) Detection and imaging of nitric oxide with novel fluorescent indicators: diaminofluoresceins. Anal Chem 70:2446–2453

6. Kojima H, Sakurai K, Kikuchi K et al (1998) Development of a fluorescent indicator for nitric oxide based on the fluorescein chromophore. Chem Pharm Bull (Tokyo) 46:373–375

7. Planchet E, Kaiser WM (2006) Nitric oxide (NO) detection by DAF fluorescence and chemiluminescence: a comparison using abiotic and biotic NO sources. J Exp Bot 57:3043–3055

8. Mur LAJ, Mandon J, Cristescu SM et al (2011) Methods of nitric oxide detection in plants: a commentary. Plant Sci 181:509–519

9. Desikan R, Cheung MK, Bright J et al (2004) ABA, hydrogen peroxide and nitric oxide signalling in stomatal guard cells. J Exp Bot 55:205–212

10. Neill S, Barros R, Bright J et al (2008) Nitric oxide, stomatal closure, and abiotic stress. J Exp Bot 59:165–176

11. García-Mata C, Lamattina L (2013) Gasotransmitters are emerging as new guard cell signaling molecules and regulators of leaf gas exchange. Plant Sci 201–202:66–73

12. Kolla VA, Raghavendra AS (2007) Nitric oxide is a signaling intermediate during bicarbonate-induced stomatal closure in *Pisum sativum*. Physiol Plant 130:91–98

13. Srivastava N, Gonugunta VK, Puli MR et al (2009) Nitric oxide production occurs downstream of reactive oxygen species in guard cells during stomatal closure induced by chitosan in abaxial epidermis of *Pisum sativum*. Planta 229:757–765

14. Gonugunta VK, Srivastava N, Puli MR et al (2008) Nitric oxide production occurs after cytosolic alkalization during stomatal closure induced by abscisic acid. Plant Cell Environ 31:1717–1724

Chapter 6

Methods to Detect Nitric Oxide in Plants: Are DAFs Really Measuring NO?

Stefan Ruemer, Markus Krischke, Agnes Fekete, Maria Lesch, Marin J. Mueller, and Werner M. Kaiser

Abstract

Nitric oxide, a gaseous radical molecule, appears involved in many reactions in all living organisms. Fluorescent dyes like DAF-2 and related compounds are still widely used to monitor NO production inside or outside cells, although doubts about their specificity have recently been raised. We present evidence that DAF dyes do not only react with nitric oxide but also with peroxidase enzyme and hydrogen peroxide. Both are secreted in the case of elicitation of tobacco suspension cells with cryptogein, with a fluorescence increase mimicking NO release from cells. However, HPLC separation shows that fluorescence outside cells does not at all originate from DAF-2T, the product of DAF-2 and NO, but from other yet unidentified compounds. Inside cells, other DAF molecules are formed but only a minor part is DAF-2T. The chemical nature of the novel DAF derivatives still needs to be determined.

Key words Nitric oxide, Cryptogein, Diaminofluorescein, Nitrite, Nitrate reductase, Nitric oxide synthase

1 Introduction

Presently, NO appears to be involved in a plethora of physiological reactions in plants, animals, and humans. One of the best-known effects is vasodilatation of human blood vessels. In plants, it has been reported to play a role in important reactions to environmental stresses, such as, e.g., stomatal movement [1]. NO has been reported to interact with the classical plant hormones in many ways. It has been named a nontraditional plant growth regulator [2].

The chemical nature of NO as a gaseous diatomic radical with an unpaired electron is responsible for its reactivity, which is also reflected by its short physical half-life which is a maximum of 500 s [3]. In aqueous solution, it is rapidly oxidized to the acidic anions nitrite (NO_2^-) and nitrate (NO_3^-). In biological surroundings containing proteins or DNA, and at NO concentrations in the μmolar range, the half-life is lowered to seconds or even milliseconds.

Kapuganti Jagadis Gupta (ed.), *Plant Nitric Oxide: Methods and Protocols*, Methods in Molecular Biology, vol. 1424, DOI 10.1007/978-1-4939-3600-7_6, © Springer Science+Business Media New York 2016

Synthesis routes for NO are debated widely. In animal cells, synthesis of NO occurs via the well-characterized reaction of l-arginine via catalysis of the NO-synthase (NOS) enzyme family, in plant cells reduction of nitrite is presently considered as the major pathway for NO synthesis. This reduction can either take place in the mitochondria of root cells using electrons from mitochondrial electron transport [4] or in a side reaction of the cytosolic enzyme nitrate reductase [5]. The existence of a plasma membrane bound nitrite:NO:oxidoreductase in root cells was also described [6]. The existence of "the" plant NOS was hypothesized and discussed extensively but its existence was never proven [7]. Recently, an oxidative pathway using hydroxylamines as a substrate was proposed but its role in vivo is still not established [8].

Effects of nitric oxide were observed or generated under conditions where its concentrations were not really known or could only be estimated. In animal cells, concentrations of NO according to [9] vary from 100 pM to a maximum of 5 nM. Probably, diffusional pathways of NO are short, and thus the effects of NO are restricted to its sites of production or to cells in the very near surrounding [10].

2   Measuring NO: Chemiluminescence vs. Fluorescent Dyes

Quantification of NO is hampered by its reactivity and short half-time. Various methods have been proposed and used to detect NO inside or outside living organisms, each with its own advantages and disadvantages [11]. One important difference between the methods does not relate to the chemical reactions involved but to whether NO from an aqueous phase (cells) has to diffuse into a gas phase to be carried (as gas) to a detector, or whether an indicator molecule (e.g., a fluorescing probe) enters the close surroundings of cells or tissues or diffuses into living cells and organelles to react with NO close to its production sites.

3   Gas Phase NO Measurement

As an example for a method based on gas phase NO measurements, a device for ozone-based chemiluminescence is depicted in the scheme of Fig. 1. It comprises the flow of NO from the sample (leaves or suspension cells) into a gas stream (air or nitrogen) carrying NO into the analyzer where it reacts with ozone to produce nitrogen dioxide in an excited state (NO_2^*) which emits light when returning to the ground state (NO_2) (Fig. 1).

In the case of gas phase measurements, it is usually not known what percentage of NO produced inside cells or tissues really arrives at the detector. On the other hand, gas phase measurements like ozone chemiluminescence have the advantage of being highly sensitive and NO specific.

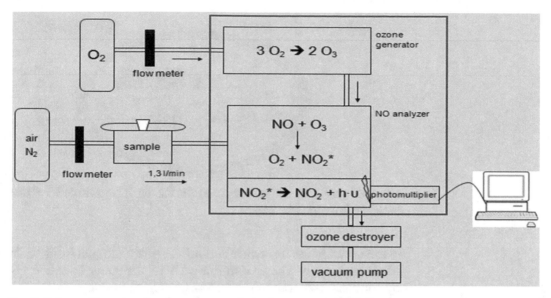

Fig. 1 Scheme of a chemiluminescence analyzer for measuring NO in a gas stream with the relevant reactions

4   Fluorescing Dyes as NO Indicators

The most common method to detect and visualize NO outside or inside living cells closer to its production sites is by fluorescent dyes on fluorescein or rhodamine base, e.g., DAF-2, DAF-FM, DAR-4M, and by their cell-permeable derivates (Fig. **2**).

Fluorescent dyes were and are extensively used to evaluate the physiological relevance of nitric oxide in medicine, botany, and zoology. Using DAF, NO has been detected, e.g., in human myometrial tissue [12] leading to its proposed role in pregnancy [13]. In botanical research, DAFs have been used to investigate the role of nitric oxide in lignification of xylem vessels [14], in opening and closure of stomatal guard cells [15], growth of the pollen tube [16] or orientation and development of roots [17].

Because of this wide spread use, documented by more than 1000 publications during the last decade in botany, medicine, or zoology, a discussion of this method and of its potential pitfalls will be given below. Consequences for the interpretation of results using these dyes will be considered.

The expected basic reaction of fluorescence indicators based on fluorescein is the nitrosation by an oxidation product of NO (probably N_2O_3 or NO^+), but not with NO itself [18], forming the highly fluorescing triazol (DAF-2T, Fig. **2**).

For in situ measurements inside living cells or tissues, these can be loaded with the diacetate derivatives of respective fluoresceins. After hydrolysis of the ester bonds inside cells, the expectation is that "NO" reacts with the dye forming again DAF-2T which is

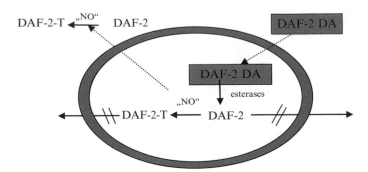

Fig. 2 Cell-permeable and noncell-permeable fluorescent dyes to detect NO and other reactive molecules (NO)

plasma membrane impermeable and therefore trapped inside cells. Measurements can be carried out with a fluorimeter in extracts or by confocal laser scanning microscopy (LSM) in samples.

The disadvantage of long diffusional pathways for NO is only marginally important in the case of fluorescence indicators. This should basically avoid large losses of NO through competing reactions with biological materials and therefore theoretically give more direct readings of NO concentrations and production rates. However, most if not all optical NO indicators seem to be subjects to side reactions, as will be shown below.

Under normal conditions, but even more in stress situations, fluorescent dyes are exposed to a variety of reactive compounds produced by cells. For example, all kinds of reactive oxygen species (ROS) may interact with DAFs. It seems inevitable that DAFs are subject to diverse chemical reactions, aside of their reaction with NO. That has been shown in a number of cases:

1. Calcium ions as well as continued exposure to light increased the NO-dependent fluorescence signal by a factor of up to 200 [19].

2. Ascorbic acid and dehydroascorbate lead to the formation of adducts showing the same fluorescence pattern but having different chemical structures than DAF-2T [20].

3. [21] showed a fluorescence increase of DAF and related NO indicators even in the complete absence of any NO.

4. DAF-2 is oxidized to a *nonfluorescent* radical derivative in the presence of reactive oxygen species that directly reacts with NO to DAF-2T [22].

Furthermore, [23] described serious discrepancies between NO measurements obtained with either gas phase chemiluminescence or DAF fluorescence using cryptogein-elicited tobacco suspension cells: While chemiluminescence started to indicate NO emission about 3 h after exposure to the elicitor, fluorescence increase could be observed only until 3–6 h [23].

5   DAFs also React with Reactive Oxygen Species (ROS) and Peroxidase

The above results prompted us to further investigate the possibility of an NO-independent formation of fluorescing DAF derivatives. The protein cryptogein from the oomycete *Phytophtora cryptogea* is a very potent activator of the hypersensitive response (HR) in tobacco, causing cell death (necrosis) within several hours [24] Importantly, NO production has been considered an essential part of the signaling chain leading to those responses. Accordingly, cryptogein was used to trigger NO production of cells, to be detected by DAF fluorescence.

In our experiments, DAF-2 fluorescence was first followed outside cells, i.e., in the growth medium of elicited and nonelicited tobacco suspension cells, eventually indicating NO release from cells. Unexpectedly, direct addition of DAF to the cell suspension gave only low and variable fluorescence increases, for as yet unknown reasons. Additionally, there was no difference between the cells in the presence or absence of cryptogein (Fig. 3).

Therefore, the experimental approach was modified: An aliquot of cells was elicited for 2 h, filtered to separate cells from the medium, and DAF was added to the clear filtrate. Subsequently fluorescence was continuously followed in a fluorimeter: When cells were elicited for about 2 h, a nearly constant fluorescence increase after addition of DAF-2 to the cell filtrate took place for about 1 h (Fig. 4). Surprisingly when a filtrate was left for some hours at room temperature, and DAF was subsequently added,

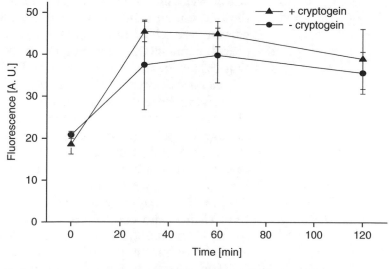

Fig. 3 Fluorescence development after addition of DAF-2 to tobacco suspension cells in the presence or absence of cryptogein (50 nM). At the indicated times, an aliquot of cells was filtered and the fluorescence in the clear filtrate was measured

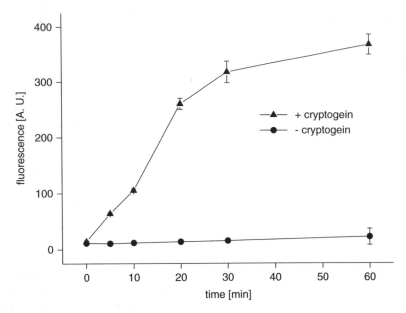

Fig. 4 Fluorescence increases after addition of DAF2 to the filtrate of elicted (+50 nM cryptogein) or nonelicited (cryptogein) tobacco suspension cells

fluorescence still increased. That was considered a strong hint that the fluorescence increase was not caused by formation of the NO reaction product DAF-2T, because after such a long time NO should have been decomposed almost completely.

Fluorescence increase with DAF-2 (also observed with DAF-FM and DAR-4M; not shown) was lower or completely absent when an inhibitor of NADPH oxidase, DPI was added to the cells during incubation or when catalase, an H_2O_2-degrading enzyme, was added to the filtrate. Apparently, other compounds, probably some ROS, secreted into the medium of tobacco suspension cells were also able to provoke DAF fluorescence, probably not via nitrosation of DAF-2, but via some unknown reaction.

Among the molecules frequently secreted by cells into the apoplast in stress situations are enzymes like peroxidase [25] and reactive oxygen species (ROS) like hydrogen peroxide or the superoxide anion. ROS and peroxidase together form lignin polymers in an oxidative cross-linking reaction that strengthen the cell wall [26]. We found both, peroxidase activity and hydrogen peroxide in cell filtrates. However, while peroxidase activity was similar in filtrates from elicited and control cells, hydrogen peroxide content was largely elevated only in the filtrate of elicited cells consistent with an oxidative burst as part of the response to the elicitor (Table 1).

The increase of DAF fluorescence was clearly reduced when potassium cyanide, a potent inhibitor of transition–metal containing enzymes, was added to the filtrate and completely absent when catalase, an H_2O_2-degrading enzyme, was added [27]. These findings suggested an NO-independent formation of fluorescing DAF

derivatives. Reverse phase HPLC separation (RP-HPLC) of the
reaction products indeed revealed that those fluorescing derivatives were different from DAF-2T (Fig. 5a), the reaction product

Table 1

Peroxidase (PO) activity and H_2O_2 content in the filtrate of suspension cells treated with or without cryptogein (50 nM) in the presence or absence of the NOX-inhibitor DPI (50 μM)

	Control (−cryptogein)	+Cryptogein
	PO activity [mU/mL cell filtrate]	
Cell filtrate	20.0±7.1	23.8±8.9
	H_2O_2 concentration [μM]	
Cell filtrate	2.5±2.2	10.4±2.2
Cell filtrate of cells + DPI	0.8±0.1	0.5±0.1
Cell filtrate of cells + cPTIO	1.0±0.3	8.4±0.3

After incubation, cells were filtered, PO activity and H_2O_2 were determined. Values given are means of eight-independent measurements ± S. D

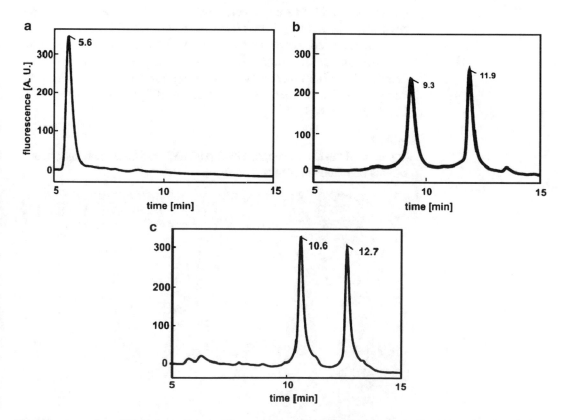

Fig. 5 Reverse phase HPLC separation and fluorescence detection of authentic DAF-2T (**a**) and of DAF-2 reaction products formed during a 1 h incubation of DAF-2 (0.5 μM) with the filtrate of nitrate-grown and cryptogein-elicited (2 h) tobacco cells (**b**) or in vitro (**c**) by incubation of horseradish peroxidase (HRPO, 0.1 U/mL) and H_2O_2 (100 μM). Numbers at the elution profiles give retention times (min)

of DAF-2 with oxidized NO. Instead, two novel reaction products with higher retention times (9 and 12 min, Fig. **5**b) were found. The same fluorescent reaction products were also found in RP-HPLC when commercially available horseradish peroxidase (HRPO) and hydrogen peroxide were reacted with DAF-2 in a cell-free medium in vitro (Fig. **5**c).

6   Influence of the NO Scavenger cPTIO

2-(4-Carboxyphenyl)-4,5-dihydro-4,4,5,5-tetramethyl-1H-imidazolyl-1-oxy-3-oxide (cPTIO) is frequently used to prove the participation of nitric oxide in physiological reactions. If cPTIO is added and the reaction, e.g., stomatal movement cannot be observed any more, and if, at the same time, a DAF fluorescence increase is prevented, both the physiological reaction and DAF fluorescence are considered NO-dependent.

We also observed the influence of cPTIO on the NO-independent formation of DAF reaction products. First, cPTIO was tested with the cell filtrate of cryptogein-treated cells (Table 2a). Fluorescence remained nearly the same level; only when cPTIO was already present during incubation it was lowered to about 30 %.

When added to an in vitro assay of HRPO + H_2O_2 + DAF fluorescence increased drastically in the complete absence of NO (Table 2b). This increase was also registered in vitro in the presence of NO-gas indicating a reaction of cPTIO with NO forming DAF-reactive N_2O_3 [28].

Table 2
Influence of the NO scavenger cPTIO (100 μM) on DAF-2 fluorescence in various situations in vitro and in vivo

	Fluorescence after 1 h [A. U.]
(a) Filtrate of elicited cells	
+DAF	672.5 ± 159.7
+cPTIO, +DAF	666.0 ± 48.8
(filtrate of elicited cells + cPTIO) + DAF	201.7 ± 9.4
(b) In vitro	
DAF + HRPO + H_2O_2 (100 μM)	397.7 ± 54.8
DAF + HRPO + H_2O_2 + cPTIO	1133.5 ± 21.2
(c) In vitro	
DAF + NO-gas (5 ppm)	3.8 ± 0.7
DAF + NO-gas + cPTIO	220.2 ± 21.2

Due to the strong blue color of the cPTIO solution, fluorescence values were corrected by 20 %

These results point to a possible reaction of cPTIO with ROS, as the level of H_2O_2 is also reduced when cPTIO is present during incubation of suspension cells (Table 1). Accordingly, the use of cPTIO for proving a participation of NO in DAF fluorescence and in physiological reactions seems also questionable.

7 Fluorescence Development Inside Cells

In another experiment, tobacco suspension cells were incubated for half an hour with DAF-2 DA, the cell-permeable diacetate derivative. After removal of excess dye and 2 h incubation with the elicitor, extraction was performed by freeze-thawing in liquid nitrogen. After centrifugation the extract of the cells was fluorimetrically analyzed. Surprisingly, even the extract of nonelicited cells showed very high background fluorescence. It was hardly elevated in the filtrate of cryptogein-elicited cells. On the other hand after application of H_2O_2, a substance supposed to increase NO production [29], fluorescence of the extract was about three times higher compared to the control.

HPLC analysis showed that bulk fluorescence inside cells originated from a group of DAF derivatives completely different from those found in the filtrate. The main part of background fluorescence in control cells was caused by a group of very early eluting hydrophilic compounds (Fig. 6a). After applying elicitor or H_2O_2,

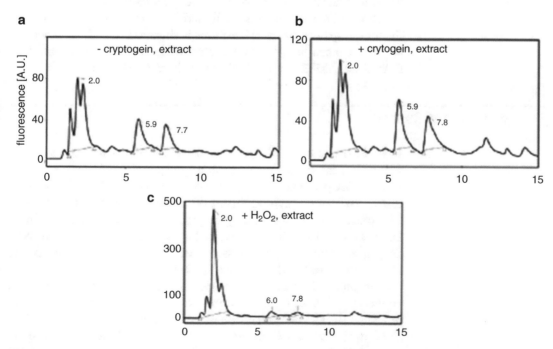

Fig. 6 Reverse phase HPLC separation of extracts from tobacco suspension cells preloaded with DAF-2 DA and incubated for 1 h without (**a**) 50 nM cryptogein (**b**), 1 mM H_2O_2 (**c**), Numbers at the peaks indicate retention time. Note the different scale (*y*-axis)

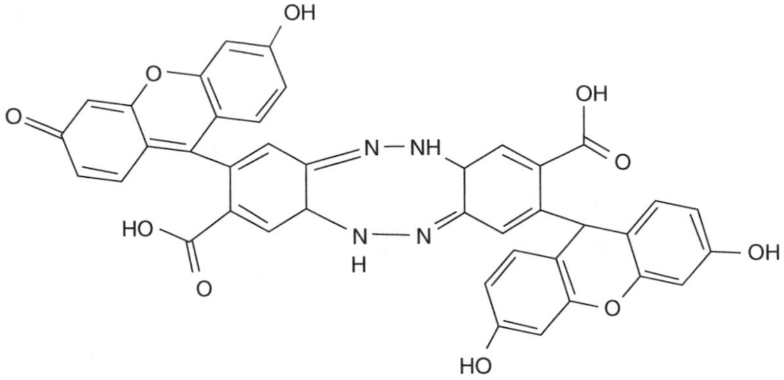

Fig. 7 Possible chemical structure of the dimer formed by two DAF-2 molecules in an in vitro reaction (with permission of Dr. Agnes Fekete, Pharmaceutical Biology, Uni Würzburg)

a peak at RT = 8 min (Fig. **6**b) or a peak in the group of early eluting DAF derivatives (Fig. **6**c) was elevated, respectively. Only a small amount of DAF-2T could be detected in the extracts.

The chemical structure of the newly synthesized fluorescing DAF reaction products has not yet been established. However, preliminary mass spectrometric analysis of an in vitro assay $(HRPO + H_2O_2 + DAF-2)$ suggest that these compounds are a dimer of DAF produced by an oxidation process with hydrogen peroxide catalyzed by peroxidase enzyme which would also match its biological reaction. A possible chemical formula of the dimer in which two DAF molecules are linked by a tetrazol ring formed by the amino groups of the monomers is shown in Fig. 7. This would fit with a more lipophilic character of the products compared to DAF-2 or DAF-2T when looking at the retention times.

8   Conclusion and Outlook

DAF-2T was not at all detected in the filtrate of tobacco suspension cells treated with the elicitor cryptogein. Even in extracts from DAF-2 DA-pretreated cells, the contribution of DAF-2T fluorescence to total cryptogein-induced fluorescence was minor. Therefore, qualitative and quantitative NO detection by those fluorescence dyes should always be accompanied by other NO detection techniques in order to avoid misinterpretations. Both, DAF dyes as well as the so-called NO-scavenger cPTIO should not be considered NO specific. In order to reliably detect nitric oxide emission, other more specific methods like gas phase chemiluminescence should be used. At least, a positive result obtained using DAF dyes should be verified by applying HPLCg separation to unravel the chemical nature of the DAF products.

More accurate analysis, e.g., of cell extracts have to be performed to elucidate the various reaction products of DAF-2 and their exact chemical structure. This would possibly involve purification procedures to remove compounds present in the extracts of cells which disturb the measurements. HPLC results suggest that inside and outside cells DAF-2 reacts with quite different metabolites in probably different reactions. Further analysis could contribute to a better understanding of the chemical reactions of fluorescence indicators in a complex cellular environment.

References

1. Garcia Mata C, Lamattina L (2001) Nitric oxide induces stomatal closure and enhances the adaptive plant responses against drought stress. Plant Physiol 126:1196–1204

2. Belgini MV, Lamattina L (2001) Nitric oxide: a nontraditional regulator of plant growth. Trends Plant Sci 6:508–509

3. Wink DA, Derbyshire JF, Nims RW, Saavedra JE, Ford PC (1993) Reactions of the bioregulatory agent nitric oxide in oxygenated aqueous media: determination of the kinetics for oxidation and nitrosation by intermediates generated in the NO/O_2 reaction. Chem Res Toxicol 6:23–27

4. Gupta KJ, Stoimenova M, Kaiser WM (2005) In higher plants, only root mitochondria, but not leaf mitochondria reduce nitrite to NO, *in vitro* and *in situ*. J Exp Bot 56:2601–2609

5. Yamasaki H, Sakihama Y (2000) Simultaneous production of nitric oxide and peroxynitrite by plant nitrate reductase: *in vitro* evidence for the NR-dependent formation of active nitrogen species. FEBS Lett 468:89–92

6. Stöhr C, Strube F, Marx G, Ullrich WR, Rockel P (2000) A plasma membrane-bound enzyme of tobacco roots catalyses the formation of nitric oxide from nitrite. Planta 212:835–841

7. Fröhlich A, Durner J (2011) The hunt for plant nitric oxide synthase (NOS): is one really needed? Plant Sci 181:401–404

8. Rümer S, Kapuganti JG, Kaiser WM (2009) Plant cells oxidize hydroxylamines to NO. J Exp Bot 60:2065–2072

9. Hall CN, Garthwaite J (2009) What is the real physiological NO concentration *in vivo*? Nitric Oxide 21:92–103

10. Malinski T, Taha Z, Grunfeld S, Patton S, Kapturczak M, Tomboulian P (1993) Diffusion of nitric oxide in the aorta wall monitored *in situ* by porphyrinic microsensors.

Biochem Biophys Res Commun 193:1076–1082

11. Mur LAJ, Mandon J, Cristescu SM, Harren FJM, Prats E (2011) Methods of nitric oxide detection in plants: a commentary. Plant Sci 181:509–519

12. Schwendemann J, Sehringer B, Noethling C, Zahradnik HP, Schaefer WR (2008) Nitric oxide detection by DAF (diaminofluorescein) fluorescence in human myometrial tissue. Gynecol Endocrinol 24:306–331

13. Sladek SM, Magness RR, Conrad KP (1997) Nitric oxide and pregnancy. Am J Physiol 272:R441–R463

14. Gabaldón C, Roos LVG, Pedreño MA, Barceló RA (2005) Nitric oxide production by the differentiating xylem of *Zinnia elegans*. New Phytol 165:121–130

15. Desikan R, Cheung MK, Bright J, Henson D, Hancock JT, Neill SJ (2004) ABA, hydrogen peroxide and nitric oxide signalling in stomatal guard cells. J Exp Bot 55:205–212

16. Prado AM, Porterfield DM, Feijo JA (2004) Nitric oxide is involved in growth regulation and re-orientation of pollen tubes. Development 11:2707–2714

17. Correa-Aragunde N, Graziano M, Lamattina L (2004) Nitric oxide plays a central role in determining lateral root development in tomato. Planta 218:900–905

18. Kojima H, Sakurai K, Kikuchi K, Kawahara S, Kirino Y, Nagoshi H, Hirata Y, Nagano T (1998) Development of a fluorescent indicator for nitric oxide based on the fluorescein chromophore. Chem Pharm Bull (Tokyo) 46:373–375

19. Broillet MC, Randin O, Chatton YC (2001) Photoactivation and calcium sensitivity of the fluorescent NO indicator 4,5-diaminofluorescein (DAF-2): implications for cellular NO imaging. FEBS Lett 491:227–232

20. Zhang X, Kim W-S, Hatcher N, Potgieter K, Moroz LL, Gillette R, Sweedler JV (2002) Interfering with nitric oxide measurements – 4,5-Diaminofluorescein reacts with dehydro-ascorbic acid and ascorbic acid. J Biol Chem 277:48472–48478

21. Gan N, Hondou T, Miyata H (2013) Spontaneous increases in the fluorescence of 4,5-diaminofluorescein and its analogs: their impact on the fluorometry of nitric oxide production in endothelial cells. Biol Pharm Bull (TOKYO) 35:1454–1459

22. Jourd'heuil A (2002) Increased nitric oxide-dependent nitrosylation of 4,5-diaminofluoresceins by oxidants: implications for the measurement of intracellular nitric oxide. Free Radic Biol Med 33:676–684

23. Planchet E, Sonoda M, Zeier J, Kaiser WM (2006) Nitric oxide (NO) as an intermediate in the cryptogein induced hypersensitive response – a critical re-evaluation. Plant Cell Environ 29:59–69

24. Ricci P, Bonnet P, Huet JC, Sallantin M, Beauvais-Cante F, Bruneteau M, Billard V, Michel G, Pernollet JC (1989) Structure and activity of proteins from pathogenic fungi Phytophthora eliciting necrosis and acquired resistance in tobacco. Eur J Biochem 183:555–563

25. Minibayeva F, Kolesnikov O, Chasov A, Beckett RP, Lüthje A, Vylegzhanina N, Buck F, Böttger M (2009) Wound-induced apoplastic peroxidase activities: their roles in the production and detoxification of reactive oxygen species. Plant Cell Environ 32:497–508

26. Veitch NC (2004) Horseradish peroxidase: a modern review of a classic enzyme. Phytochemistry 65:249–259

27. Rümer S, Krischke M, Fekete A, Müller MJ, Kaiser WM (2012) DAF-fluorescence without NO: elicitor treated tobacco cells produce fluorescing DAF-derivatives not related to DAF-2 triazol. Nitric Oxide 27:123–135

28. Akaike T, Maeda H (1996) Quantitation of nitric oxide using 2-phenyl-4,4,5,5-tetramethylimidazoline-1-oxyl 3-oxide (PTIO). Methods Enzymol 268:211–222

29. Lum HK, Butt YKC, Sc L (2002) Hydrogen peroxide induces a rapid production of nitric oxide in mung bean (Phaseolus aureus). Nitric Oxide 6:205–213

Chapter 7

A Novel Protocol for Detection of Nitric Oxide in Plants

Prachi Jain, Anisha David, and Satish C. Bhatla

Abstract

Detection of nitric oxide (NO) in plant cells is mostly undertaken using diaminofluorescein (DAF) dyes. Serious drawbacks and limitations have been identified in methods using DAF as a probe for NO detection. The present work reporting an alternative fluorescent probe for NO detection is thus proposed for varied applications in plant systems for physiological investigations. This method involves a simple, two-step synthesis, characterization, and application of MNIP-Cu {Copper derivative of [4-methoxy-2-(1H-napthol[2,3-d]imidazol-2-yl)phenol]} for specific and rapid binding with NO, leading to its detection in plant cells by epifluorescence microscopy and confocal laser scanning microscopy (CLSM). Using sunflower (*Helianthus annuus* L.) whole seedlings, hypocotyl segments, stigmas from capitulum, protoplasts, and isolated oil bodies, present investigations demonstrate the versatile nature of MNIP-Cu in applications for NO localization studies. MNIP-Cu can detect NO in vivo without any time lag (ex. 330–385 nm; em. 420–500 nm). It exhibits fluorescence both under anoxic and oxygen-rich conditions. This probe is specific to NO, which enhances its fluorescence due to MNIP-Cu complexing with NO and treatment with PTIO leads to quenching of fluorescence. It is relatively nontoxic when used at a concentration of up to 50 μM.

Key words Confocal laser scanning microscopy (CLSM), Copper derivative of 4-methoxy-2-(1H-napthol[2,3-d]imidazol-2-yl)phenol (MNIP-Cu), Diaminofluorescein (DAF), Epifluorescence microscopy, In vivo localization, Nitric oxide (NO), Hypocotyl, Oil bodies, Protoplasts, Stigma papillae, Sunflower, Whole seedlings

1 Introduction

Most of the fluorescent probes being currently used for detection of cellular nitric oxide (NO) do not bind directly to NO, but to its other reactive forms, like N_2O_3. So far, fluorescent probes like acridine-TEMPO-DTCS-Fe(II), CuFl, FNOCT, and MNIP-Cu have been reported to bind directly with NO [1, 2]. Thus, they are highly specific probes for NO detection. Diaminofluorescein (DAF) has been extensively used for detection of NO in plants. Fluorescence due to DAF is generated by the following mechanism:

Kapuganti Jagadis Gupta (ed.), *Plant Nitric Oxide: Methods and Protocols*, Methods in Molecular Biology, vol. 1424, DOI 10.1007/978-1-4939-3600-7_7, © Springer Science+Business Media New York 2016

$$\text{DAF-2DA} \xrightarrow{\text{Esterase activity}} \text{DAF-2} \xrightarrow{\text{autoxidation of NO to } N_2O_3} \text{DAF-2T}$$

DAF-2DA (non-fluorescent) fluorescence)   DAF-2 (in presence of O_2 or NO_2)   DAF-2T (green $\lambda ex=485$ nm, $\lambda em=515$nm

This chemical transformation is based on the reactivity of two electron donating aromatic amines (attached to fluorescein) with NO in the presence of O_2 (less electron donating groups), thereby forming a highly fluorescent triazole ring. A number of serious drawbacks have been identified in the detection of NO using DAF as a fluorescent probe. Specifically, these are as follows:

1. DAF does not detect NO directly. It detects its oxidized form, i.e., N_2O_3, under oxygenic conditions [3–5]. DAF cannot be used to image NO under hypoxic or anoxic conditions due to oxygen requirement. Oxidation of NO to N_2O_3 is a third order reaction and is likely to be a rate limiting step in the formation of diaminofluorescein-2 triazole (DAF-2T), thus leading to very slow formation of N_2O_3 at nanomolar concentrations of NO, consequently affecting the real time detection of NO [5, 6].

2. ROS are by-products of metabolism but under stress conditions, increased production of ROS takes place; hence, the following reaction occurs when cells/tissues are incubated with DAF [4]:

$$\text{DAF-2} \longrightarrow \text{DAF (free radical)}$$
directly combines with NO to form DAF-2T

$$\text{NO} \xrightarrow{\text{ROS}} N_2O_3 \text{(combines with DAF-2 and forms DAF-2T)}$$

3. Formation of DAF-2T is sensitive to pH. N_2O_3 is susceptible to nucleophilic attack and OH⁻, being a strong nucleophile, competes with DAF, thus leading to a reduction in the formation of DAF-2T at pH above 7.0 [5].

4. Irrespective of the type of buffer used, N_2O_3 generated DAF fluorescence also depends on the ionic concentration of the buffer. At higher ionic strength, a reduction in DAF fluorescence has been observed. Differential loading of DAF dyes into different tissues has also been reported.

5. Ascorbic acid (AA)/dehydroascorbic acid (DHA) [generated during stress] also combine with DAF to form DAF-2-DHAs, and generate fluorescence similar to DAF-2T. According to Stöhr and Stremlau [7], the fluorescence observed in root hair and in the apical cells of root tips is primarily due to the accumulation of AA and DHA. A zonal gradation of AA and DHA in roots has been reported, with their maximum accumulation in the tip region.

In the present work, application of MNIP-Cu [Cu derivative of 4-methoxy-2-(1H-napthol[2,3-*d*]imidazol-2-yl)phenol] as a novel probe for NO detection in plant systems is presented. NO signal has been imaged in whole seedlings, hypocotyl segments exhibiting adventitious rooting, stigma papillae, protoplasts, and oil bodies. These observations suggest the utility of MNIP-Cu as a substitute for DAF for NO imaging over a wide range of tissues, cells and at subcellular levels, thereby providing a NO-specific alternative probe for varied applications in plant biology. MNIP-Cu has earlier been exploited for detection of NO signal in macrophages and endothelial cells in animal system [8].

MNIP-Cu (non-fluorescent)

Blue fluorescence
λex. =330-385 nM, λem. = 420 nM

MNIP-Cu is nontoxic at the concentrations applicable for its use (10–50 μM), is cell permeable, rapidly binds to NO directly at its intracellular sites, and specificity of fluorescence due to MNIP-NO complex is evident from its quenching by PTIO (a well-known NO scavenger). In contrast with DAF, which binds with N_2O_3 (and not NO), MNIP-Cu complexes with NO directly. The present work reports about the synthesis, characterization, and application of MNIP-Cu for specific detection of NO in different plant tissues, protoplasts, and subcellular compartments.

2 Materials

2.1 Solutions/Reagents

1. 2-hydroxy-4-methoxybenzaldehyde.
2. 3-aminonaphthalene.
3. Nitrobenzene.
4. *n*-hexane.
5. Diethyl ether.
6. *d*6-DMSO.
7. Tetramethylsilane (TMS).
8. Dimethyl sulfoxide (DMSO).
9. 10 mM: MNIP stock Store at –20 °C.
10. 50 mM: Aqueous copper sulfate.
11. 50 μM: MNIP-Cu.
12. 10 mM, pH-7.4: Tris buffer:
13. 1 mM: cPTIO.

2.2 Equipment	1. NMR Spectrometer (Delta, JEOL, Japan).
	2. 2000 FT-IR (Perkin-Elmer, USA).
	3. Confocal laser scanning microscope (TCS SP2, Leica, Germany).
	4. EC3 imaging system (UVP, UK).
	5. Epifluorescence microscope (Axioskop, Zeiss, Germany).
2.3 Plant Materials	1. 2-day-old, dark-grown sunflower seedlings grown in the absence or presence of 120 mM NaCl.
	2. Hypocotyl explants exhibiting adventitious rooting upon being subjected to 10 μM IAA treatment.
	3. Mature stigma from sunflower capitulum.
	4. Enzymatically isolated protoplasts from hypocotyls of 4-day-old, light-grown sunflower seedlings.
	5. Oil bodies isolated from cotyledons of 2-day-old, dark-grown sunflower seedlings.

3 Methods

Carry out all steps in dark at room temperature unless otherwise specified.

3.1 Synthesis and Characterization of MNIP

MNIP [4-methoxy-2-(1H-napthol[2,3-*d*]imidazol-2-yl)phenol] was synthesized with minor modifications in the available procedure [8].

1. Reflux a mixture of 2-hydroxy-4-methoxybenzaldehyde [1] (0.316 mmol), 2, 3-diaminonaphthalene [2] (0.316 mmol) and nitrobenzene [3] (2 mL) for 2 h.

2. After completion of the reaction, cool the reaction mixture to room temperature and dilute with *n*-hexane.

3. Filter the precipitate of MNIP [4] so formed and wash with diethyl ether (Fig. 1) (*see* **Note 1**).

4. Characterize the MNIP thus synthesized as light brown solid (20 mg [22 %], mp 296–297 °C) by ^{1}H NMR and IR spectroscopy.

5. Record the ^{1}H NMR spectra in *d6*-DMSO, using tetramethylsilane (TMS) as internal standard (Fig. 2). IR (KBr) Vmax (cm[-1]):3356, 2926, 2373, 1596, 1466, 1399, 1353, 1305, 1261, 1208, 1173, 1136, 1028, 953, 860/cm; ^{1}H NMR (400 MHz, DMSO-*d6*) δ: 13.48 (brs, 1H, NH), 13.06 (brs, 1H, OH), 8.14 (brs, 1H, ArH), 7.99–8.04 (m, 4H, ArH), 7.37–7.40 (m, 2H, ArH), 6.61–6.64 (m, 2H, ArH), 3.82 (s, 3H, OCH3). Match the ^{1}H NMR data obtained with that reported in literature [8].

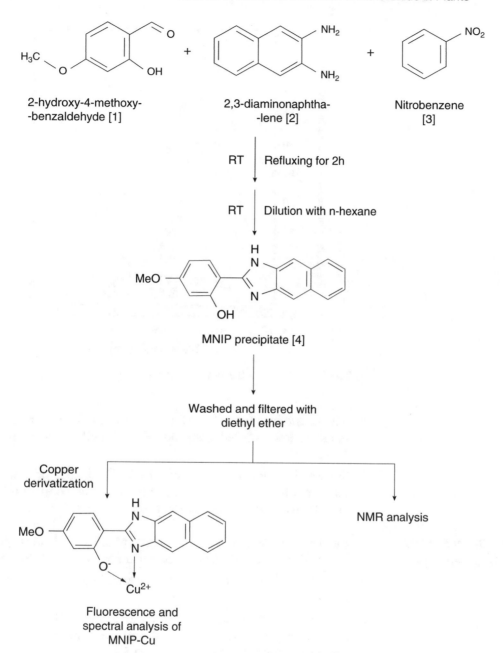

Fig. 1 Scheme of reactions leading to the synthesis of MNIP-Cu

3.2 Synthesis of MNIP-Cu

1. In order to achieve the synthesis of MNIP-Cu, dissolve the crystals of MNIP obtained in dimethyl sulfoxide (DMSO) to make a 10 mM stock (*see* **Note 2**).

2. Dilute the prepared MNIP stock (10 mM) to 1 mM with DMSO.

3. To 1 mL of 1 mM MNIP solution, add 20 μL of 50 mM of aqueous copper sulfate.

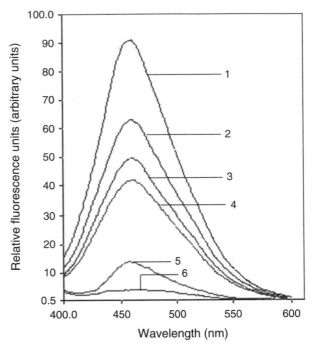

Fig. 2 NMR spectrum of MNIP synthesized according to Ouyang et al. [8]

4. Stir the mixture for at least 5 min at room temperature until a stable yellow colored solution of MNIP-Cu is formed (*see* **Note 3**).

5. Record the fluorescence spectra of MNIP in the presence of varying concentrations of copper (as aqueous copper sulfate) in order to validate its emission properties (Fig. 3).

3.3 Detection of NO in Whole Seedlings

1. Incubate the seedlings grown in the absence and presence of 120 mM NaCl in 50 μM of MNIP-Cu for 15 min (*see* **Note 4**).

2. After MNIP-Cu treatment, rinse the seedlings in distilled water for 5 min.

3. Visualize the fluorescence from tissue using UVP EC3 imaging system (ex. 385 nm; em. 420 nm) and capture the images using the attached camera (*see* **Note 5**).

Seedlings exhibit remarkable fluorescence due to NO in roots. Fluorescence is visible in the control seedlings (grown in the absence of NaCl), primarily in the elongation zone (till the transition zone), with negligible fluorescence in the root tips. Whereas seedlings stressed with 120 mM NaCl, roots show fluorescence in the differentiation and elongation zones as well as in the marginal cells of cotyledons (Fig. 4a). Thus, a gradation in NO accumulation in seedlings is evident. Similar detection of whole seedling fluorescence due to NO has recently been reported in *Arabidopsis*, using DAF as a probe [9].

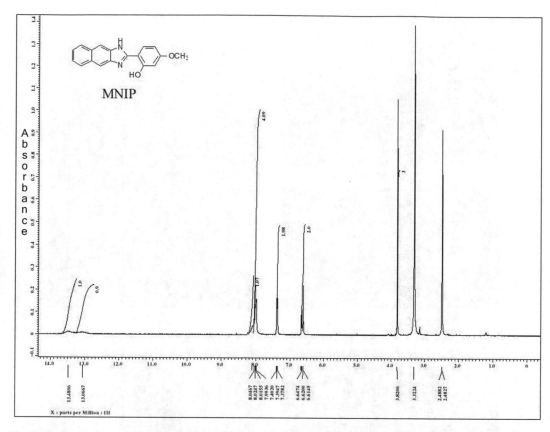

Fig. 3 Fluorescence emission spectra of MNIP (10 μM) in the absence and presence of various concentrations (0.625–10 μM) of copper (provided as CuSO₄). (*1*) MNIP (10 μM), (*2*) MNIP + 0.625 μM CuSO₄, (*3*) MNIP + 1.25 μM CuSO₄, (*4*) MNIP + 2.5 μM CuSO₄, (*5*) MNIP + 5 μM CuSO₄, (*6*) MNIP + 10 μM CuSO₄ (ex. 385 nm; em. 450 nm)

3.4 Detection of NO in Sunflower Hypocotyls Explants Exhibiting Adventitious Rooting

1. Incubate the hypocotyl explants in 50 μM MNIP-Cu and visualize fluorescence due to MNIP-NO complex within 5–10 min, using UVP EC3 imaging system (ex. 385 nm; em. 420 nm) (*see* **Note 6**).

Although some reports on NO production in adventitious roots (as evident using DAF as a probe) is available [10], so far very few attempts have been made to image NO distribution in whole explants [9]. Hypocotyl explants subjected to 10 μM IAA treatment for 4 days exhibit only internal rooting. In this case, incubation with MNIP-Cu does not show any significant fluorescence from the basal cut ends (Fig. 4b). However, 5 days after incubation, when roots just emerge from the epidermis close to the cut ends of the explants, fluorescence due to NO is clearly evident from the emerging roots (Fig. 4b). Interestingly, 7 days after incubation in IAA, root extension growth takes place and the extreme tips, representing the meristematic zone, show intense fluorescence due to NO (Fig. 4b). There is a surge in NO production till the base of the emerging roots. Thus, a differential distribution of NO in the different stages of adventitious rooting is evident [11].

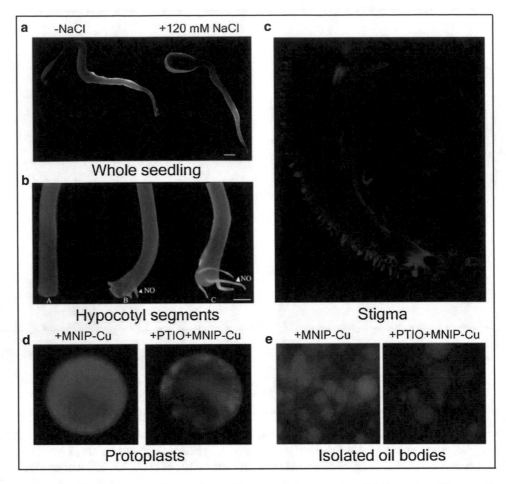

Fig. 4 Detection of NO generation in seedlings (**a**), hypocotyl segments exhibiting adventitious rooting (**b**), stigma papillae (**c**), protoplasts (**d**), and isolated oil bodies (**e**) upon treatment with MNIP-Cu. Other details are as given in "Materials" and "Methods"

3.5   Detection of NO on Stigma Surface

1. Immerse freshly plucked mature stigmas at mature stages of development in 1 ml of 50 μM of MNIP-Cu Wash them in fresh buffer for 10 min.

2. Visualize stigma surface in CLSM by undertaking z-series imaging to identify fluorescence indicative of NO presence (ex. 365 nm and em. 460 nm) in various cellular regions using Argon-UV lasers and a pin hole of one.

3. Stigma papillae exhibit intense blue fluorescence from their apoplast and cytoplasm (Fig. 4c). This observation indicates a critical role of endogenous NO in modulating sigma receptivity during pollen–stigma interaction [12].

3.6   Detection of NO in Protoplasts

1. Purified protoplast preparations are treated with 25 μM of MNIP-Cu for 5 min (*see* **Notes** 7 and 8).

2. Visualize fluorescence due to NO after exciting protoplasts at 365 nm (em. 420 nm) using epifluorescence microscope and capture images using the attached camera (*see* **Note 9**).

3. Incubate protoplasts with 1 mM PTIO (prepared in 10 mM Tris–HCl buffer, pH 7.4) for 30 min, followed by co-incubation with MNIP-Cu for 10 min.

4. Record the observations on epifluorescence microscope at 630× and capture images using the attached camera.

Blue fluorescence is evident from the cytoplasm, representing NO distribution. cPTIO treatment leads to significant reduction in fluorescence [11] (Fig. 4d).

3.7 Localization of Nitric Oxide on Isolated Oil Bodies (OBs)

1. Incubate the isolated oil bodies with 25 μM of MNIP-Cu for 10 min (*see* **Note 10**).

2. Visualize after excitation at 365 nm (em. 420 nm) using epi-fluorescence microscope (*see* **Note 11**).

3. To confirm that the fluorescence is specifically due to NO, incubate the oil body suspension in dark for 1 h at 25 °C in a solution of 1 mM cPTIO. Following this, incubate in 25 μM MNIP-Cu for another 10 min.

4. Record observations at 1000× using epifluorescence microscope.

MNIP-Cu bound to NO is intensely fluorescent on the surface of OBs. Upon cPTIO treatment, fluorescence from OBs drastically quenches, indicating the specificity of fluorescence from NO (Fig. 4e). A comparison with results obtained using DAF as a probe has revealed that detection of subcellular fluorescence on the surface of isolated OBs is achievable with greater ease using MNIP-Cu. Fluorescence is also more intense by the application of this probe (MNIP-Cu) indicating greater sensitivity of MNIP-Cu for NO. These observations indicate a signaling role of NO on OB surface during oil seed germination.

To sum up, MNIP-Cu has been found to substitute DAF as a more specific probe for localization of NO in whole seedlings, hypocotyl explants showing adventitious rooting, stigma papillae, protoplasts, and isolated oil bodies. Its synthesis is simple and application straightforward, without any adverse effect on cell viability at the concentrations used. For the first time, spatial distribution of NO has been demonstrated in whole seedlings, hypocotyl explants, and stigma papillae. Adventitious root differentiation from hypocotyls explants coincides with enhanced NO accumulation in the zones of differentiation and elongation of explants. NO accumulation is also demonstrated from experiments on protoplasts isolated from hypocotyls, isolated oil bodies, and stigma papillae. This novel procedure, thus, provides a convenient and more specific protocol to decipher NO-associated signaling mechanisms in varied aspects of plant development at whole tissue, cellular, and subcellular levels.

4 Notes

1. MNIP crystals can be stored at 25 °C until use.

2. A DMSO stock of MNIP can be stored at –20 °C for 6–8 weeks.

3. MNIP-Cu is always prepared fresh from MNIP stock just before use.

4. MNIP-Cu is found to be tolerable by plant cells at various concentrations tried (10–50 μM).

5. The range of MNIP-Cu excitation is 330–385 nm and that of emission is 420–500 nm.

6. Whole hypocotyl explants require 50 μM for optimal fluorescence detection.

7. Protoplasts remain viable for more than 3 h when incubated with 25 μM of MNIP-Cu.

8. Protoplasts are recommended to be incubated at 20–25 °C and should be subjected to MNIP-Cu as soon as possible, preferably within 30 min of isolation and washing.

9. Do not expose protoplasts directly to excitation wavelength. First select appropriate protoplasts in visible light and then expose them to excitation wavelength for NO detection and capture the image.

10. Ensure proper washing of oil bodies in the suspension buffer after MNIP-Cu treatment but prior to observations.

11. Although MNIP-Cu treated tissue/cells/protoplasts do not exhibit quenching of fluorescence, prolonged exposure to excitation wavelength should be avoided.

References

1. Ye X, Rubakhin SS, Sweedler JV (2008) Detection of nitric oxide in single cells. Analyst 133:423–433

2. Vandana S, Sustmann R, Rauen U, Stöhr C (2012) FNOCT as a fluorescent probe for *in vivo* localization of nitric oxide distribution in tobacco roots. Plant Physiol Biochem 59: 80–89

3. Kojima H, Nakatsubo N, Kikuchi K et al (1998) Detection and imaging of nitric oxide with novel fluorescent indicators: diaminofluoresceins. Anal Chem 70:2446–2453

4. Planchet E, Kaiser WM (2006) Nitric oxide production in plants: facts and fictions. Plant Signal Behav 1:46–51

5. Viteceka J, Reinohl V, Jones RL (2008) Measuring NO production by plant tissues and suspension cultured cells. Mol Plant 1:270–284

6. Mayer B, Hemmens B (1997) Biosynthesis and action of nitric oxide in mammalian cells. Trends Biochem Sci 22:477–481

7. Stöhr C, Stremlau S (2006) Formation and possible roles of nitric oxide in plant roots. J Exp Bot 57:463–470

8. Ouyang J, Hong H, Shen C et al (2008) A novel fluorescent probe for the detection of nitric oxide *in vitro* and *in vivo*. Free Radic Biol Med 45:1426–1436

9. Arita NO, Cohen MF, Tokuda G, Yamasaki H (2007) Fluorometric detection of nitric oxide with Diaminofluoresceins (DAFs): applications and limitations for plant NO research. In: Lamattina L, Polacco JC (eds) Nitric oxide in

plant growth, development and stress physiology. Springer, Berlin, Heidelberg, pp 269–280

10. Huang AX, She XP, Huang C, Song TS (2007) The dynamic distribution of NO and NADPH-diaphorase activity during IBA-induced adventitious root formation. Physiol Plant 130:240–249

11. Yadav S, David A, Baluska F, Bhatla SC (2010) Rapid auxin-induced nitric oxide accumulation and subsequent tyrosine nitration of proteins during adventitious root formation in sunflower hypocotyls. Plant Signal Behav 8:e23196

12. Sharma B, Bhatla SC (2013) Accumulation and scavenging of reactive oxygen species and nitric oxide correlate with stigma maturation and pollen-stigma interaction in sunflower. Acta Physiol Plant 35:2777–2787

Detection of Nitric Oxide by Electron Paramagnetic Resonance Spectroscopy: Spin-Trapping with Iron-Dithiocarbamates

Luisa B. Maia and José J.G. Moura

Abstract

Electron paramagnetic resonance (EPR) spectroscopy is the ideal methodology to identify radicals (detection and characterization of molecular structure) and to study their kinetics, in both simple and complex biological systems. The very low concentration and short life-time of NO and of many other radicals do not favor its direct detection and spin-traps are needed to produce a new and persistent radical that can be subsequently detected by EPR spectroscopy.

In this chapter, we present the basic concepts of EPR spectroscopy and of some spin-trapping methodologies to study NO. The "strengths and weaknesses" of iron-dithiocarbamates utilization, the NO traps of choice for the authors, are thoroughly discussed and a detailed description of the method to quantify the NO formation by molybdoenzymes is provided.

Key words Nitric oxide radical, Electron paramagnetic resonance (EPR), Spin-trap, Iron-dithiocarbamate, Nitrite, Xanthine oxidoreductase, Aldehyde oxidoreductase

Abbreviations

AOR	Aldehyde oxidoreductase
CPTIO	2-(4-Carboxyphenyl)-4,4,5,5-tetramethyl-imidazoline-1-oxyl-3-oxide
DETC	Diethyldithiocarbamate
DMPO	5,5-Dimethyl-1-pyrroline N-oxide
EPR	Electron paramagnetic resonance
Fe-$(TC)_2$	Iron-*bis*-dithiocarbamates
Fe^{2+}-$(TC)_2$	Ferrous-*bis*-dithiocarbamate
Fe^{3+}-$(TC)_2$	Ferric-*bis*-dithiocarbamate
Hb	Hemoglobin
MGD	N-methyl-D-glucamine-dithiocarbamate
MNIC	Mononitrosyl-iron complex
MNP	2-Methyl-2-nitrosopropane
NO	Nitric oxide radical ($^\bullet$NO)

Kapuganti Jagadis Gupta (ed.), *Plant Nitric Oxide: Methods and Protocols*, Methods in Molecular Biology, vol. 1424,
DOI 10.1007/978-1-4939-3600-7_8, © Springer Science+Business Media New York 2016

ST	Spin-trap molecule
TEMPO	2,2,6,6-Tetramethylpiperidinyl-N-oxyl
XO	Xanthine oxidase

1 Introduction

Nitric oxide radical (chemical formula ˙NO, abbreviated as NO), is a signaling molecule involved in several physiological processes, in mammals, plants, and prokaryotes [1–8]. However, while in some situations the participation of NO is already compelling, in many other cases its involvement is less certain. Also the sources of NO are a matter of debate. Presently, different pathways are thought to contribute to the NO generation, including mammalian and pro-karyotic NO synthases and several mammalian, plant, and prokary-otic "non-dedicated" nitrite reductases, that make use of metalloproteins, present in cells to carry out other functions, to reduce nitrite to NO. Hence, the advancement of knowledge of NO Biology demands for methods, not only to quantify, but also to unequivocally identify the NO. Several methodologies were developed to measure NO, either indirectly, quantifying for exam-ple the NO oxidation products, and directly, exploring different properties of NO, such as its ability to react with ozone to produce light (chemiluminescence) or its radical nature (electron paramag-netic resonance (EPR) spectroscopy).

NO, with its 11 valence electrons, has an unpaired electron in a π-antibonding orbital, polarized toward the nitrogen atom (the unpaired electron is delocalized over the nitrogen) (Fig. 1). As a result, NO is a radical molecule and this feature makes the EPR spec-troscopy an ideal methodology to both quantify and identify NO.

1.1 Basic EPR Theory

Because of its spin, the electron has a magnetic moment and, in the presence of an applied magnetic field (B_0), the magnetic moment has two allowed orientations (the principles of quantum mechanics states that only two energy states or levels are allowed), corre-sponding to two spin states, with energies (E),

$$E_\alpha = \tfrac{1}{2}g\mu_B B_0 \text{ and } E_\beta = -\tfrac{1}{2}g\mu_B B_0 \tag{1}$$

where μ_B is the Bohr magneton ($=9.2740154\,(31) \times 10^{-24}\,\mathrm{JT^{-1}}$) and the parameter g (also called g-factor) is a constant that is dependent on the nature, structure, and environment of the paramagnetic species (for a free electron, g = 2.0023). Transitions between the two states can be induced when electromagnetic radiation with energy equal to $\Delta E = E_\alpha - E_\beta$ is applied perpendicularly to the exter-nal magnetic field, or better saying, radiation of appropriate fre-quency, v (because $E = h v$, where h is the Planck constant ($=6.6260755\,(40) \times 10^{-34}$ Js)),

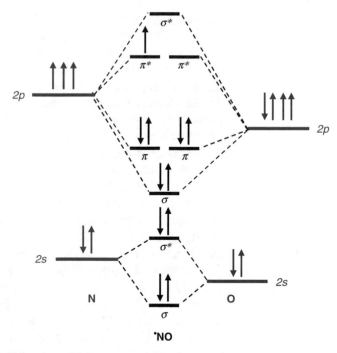

Fig. 1 Molecular orbital diagram for NO

$$hv = \Delta E = g\mu_B B_0 \qquad (2a)$$

This is the relationship (the resonance condition) that must be satisfied for microwave radiation ($\approx 10^9$–10^{11} Hz) absorption to occur. Hence, in simple words, in EPR spectroscopy one detects unpaired electrons by measuring the absorption of electromagnetic radiation, when the paramagnetic sample is placed in a magnetic field (e.g., $B_0 \approx 10$–600 mT for a spectrometer that operates at ≈ 9.5 GHz). One can also quantify the concentration of the paramagnetic species, by simple calculation of the area under the absorption line (that is, the double integral of the spectrum, because the EPR spectrum is recorded as the first derivative of absorption). Contrary to absorbance measurements, the intensity of an EPR spectrum is independent of the species that has originated it and, thus, quantification does not require the previous knowledge of an extinction coefficient.

The parameter g, which can be identified as the spectroscopic manifestation of the magnetic moment, is obtained experimentally from Eq. 2a,

$$g = (h / \mu_B)(v / B_0) = 7.14 \times 10^{-11} \text{Ts} \, (v(s^{-1}) / B_0(\text{T})) \qquad (2b)$$

It is a measure of the local magnetic and electric fields experienced by the unpaired electron(s) and it provides valuable information regarding the nature, structure, and environment of the paramagnetic species and, therefore, information about the identity of the paramagnetic species. Regarding nature of radical species, while paramagnetic transition metal ions, such as Mo^{5+}, Fe^{3+}, or Cu^{2+} (with d^1, d^5, or d^9 configuration, respectively), display a large range of g values (g$\approx$0–10), organic radicals have g values around the one of free electron (g$\approx$2). Nevertheless, the presence of electron donor or acceptor groups modify the g value (e.g., the delocalization of the unpaired electron over heteroatoms generally causes an increase in the g value) and such effects can be used as "diagnostic" of the structure of the detectable paramagnetic species. A substantial understanding of the EPR phenomenon is required to obtain structural information from this technique (which is outside the scope of this chapter, please refer to refs. 9–11). More simple and relevant in the context of this chapter, is the effect caused by the presence of neighboring nuclei with nuclear magnetic moment ($I\neq0$, e.g., $I\,(^{1}H)=1/2$, $I\,(^{14}N)=1$, or $I\,(^{15}N)=1/2$), which create a local magnetic field that is superimposed onto the applied one (B_0). In general, there will be $2I+1$ orientations for a given nuclear spin ($2I+1$ allowed energy states); the unpaired electron experiences these different magnetic fields and the EPR signal is split into $2I+1$ lines of equal intensity and equally spaced. For example, for one unpaired electron localized into a nitrogen atom ($I\,(^{14}N)=1$), the EPR signal consists of a 1:1:1 triplet of equally spaced lines. The magnitude of the splitting between the lines is called hyperfine splitting constant (if it arises from the "parent" nucleus) or super-hyperfine splitting constant (when it arises from a satellite nucleus; both abbreviated as a); equivalent nuclei give rise to equivalent splitting constant values, while nonequivalent nuclei originate different patterns of lines. In addition, the EPR spectrum reflects also the environment of the paramagnetic species (physical state, solvent) which also determine its "motion" (refer to refs. 9–11).

In summary, EPR is a powerful spectroscopy that probes the paramagnetic center, by defining the size and shape of the magnetic moment produced by the unpaired electron(s) and by characterizing the magnetic field created by the vicinity of the unpaired electron(s). It can demonstrate the presence of paramagnetic species (radicals and some transition metal ions) and provide information on its concentration, nature and structure (identity) mobility, and intra- and intermolecular interactions. Moreover, as only unpaired electrons are able to absorb microwave radiation under these conditions, the EPR spectroscopy is specific for paramagnetic species. Hence, regardless of the complexity of the sample, there will be no spectroscopic signal (absorption) apart from the one of the paramagnetic species (i.e., there will be no background signal from the biological matrix).

1.2 Spin-Trapping

In spite of its radical nature, the characteristics of NO (and of many other radicals) do not favor its direct detection by EPR spectroscopy [7, 12–17]: (1) generation in low concentration (e.g., 10^{-9}–10^{-7} M in humans), (2) ability to freely permeate membranes (k diffusion$\approx 10^{10}$–10^{11} $M^{-1}\,s^{-1}$), (3) reactivity with dioxygen ($k\approx10^6$–10^7

$M^{-2}s^{-1}$), (4) high reactivity with superoxide anion radical ($k \approx 10^9$–10^{10} $M^{-1}s^{-1}$), (5) with metalloproteins (mostly haems and labile [4Fe–4S] centers, to yield nitrosyl derivates (metal-N=O)), and (6) with cysteine residues and other thiols (to yield S-nitrosothiol derivates (–S–N=O)). Hence, the NO available to be detected (difference between NO formed and NO consumed) is usually rather low, particularly in complex biological samples, where several "sinks" of NO could be present.

As a result, with the exception of a few radicals with long half-life (e.g., ascorbyl and tocopheroxyl radicals that are intrinsically stabilized by electron delocalization and less reactive), one needs first to "stabilize" the radical to, then, be able to detect it. This "stabilization" is achieved by spin-trapping and it involves the reaction of the radical ($X^{\bullet}$) with a spin-trap molecule (ST) to yield a new and persistent radical ($^{\bullet}$ST-X; Eq. 3) that can be subsequently detected by EPR spectroscopy.

$$X^{\bullet} + ST \rightarrow {}^{\bullet}ST\ X \qquad (3)$$

In practical terms, the spin trap, in many cases a nitrone (R_1-(R_2=) N^+–O^-, e.g., DMPO) or a nitroso (R_1-N=O, e.g., MNP) compound, is added at an appropriate concentration to ensure the effective trapping of the radicals present in the sample (the trap concentration is dictated by the ratio between the rate constant of $^{\bullet}$ST-X formation and the rate constants of radical consumption reactions; the slower the $^{\bullet}$ST-X formation reaction, the higher the trap concentration should be, in order to be able to effectively compete with the other radical consumption reactions). The stable nitroxide radicals formed (R_1-(X-R_2-) N–O$^{\bullet}$ or R_1-N(-X)–O$^{\bullet}$) are, then, detected. Because different $^{\bullet}$ST-X radicals give rise to distinctive EPR signals (determined by the $X^{\bullet}$ structure), the spectra can be used to identify the radical $X^{\bullet}$, as well as, to follow the kinetics of its formation/consumption. This approach has been widely used to study not only the obvious *small* radicals, such as the superoxide anion ($O_2^{\bullet-}$) and hydroxyl (HO$^{\bullet}$) radicals [18–22], but also radicals on biomacromolecules, proteins, lipids, and nucleic acids [23–25].

To trap NO, three main types of spin-traps have been explored, nitronyl nitroxides, hemoglobin (and myoglobin), and iron-*bis*-dithiocarbamates.

Nitronyl nitroxides (e.g., CPTIO) are N-oxo-imidazoles radicals that react with NO to yield nitrogen dioxide radical ($^{\bullet}$NO$_2$) and an imino-nitroxide radical [26, 27]. Hence, there is no NO trapping, what hampers studies with labeled nitrogen (see below how convenient this could be). In addition, because both the spin-trap itself and its product (the imino-nitroxide radical) are radicals, the resulting EPR spectrum obtained after the reaction of NO with the nitronyl nitroxide is the sum of two overlapped signals, what makes the spectrum interpretation and quantification not a straightforward task [28]. Moreover, also because these spin-traps give rise to an EPR signal, one cannot increase their concentration to overcome the relatively low rate constant of the reaction of nitronyl nitroxide with NO ($k \approx 10^3$ M^{-1}s^{-1}). On the other hand, the nitrogen dioxide radical formed reacts with NO to convert it into nitrite [8, 29, 30] and, thus, leads to an undervaluation of NO concentration (in other words, the ratio between the NO and the imino-nitroxide signal is not one to one) [28]. The nitrogen dioxide radical could also compromise the integrity of the biological sample (it is a powerfully oxidizing and nitrating agent that can nitrate protein residues, fatty acids, and nucleotides) [8]. Finally, because nitroxides are good oxidants, one must evaluate if these compounds are interfering with redox chemistry of the system under study [31, 32].

Hemoglobin (Hb) (and also myoglobin) has been widely used in NO research. It is cheap, reacts with NO with high rate constants [$k > 10^7$ M^{-1}s^{-1} (*reviewed in* ref. 8)] and can be used in UV–visible [33] and EPR spectroscopy. Besides being used in detection/quantification, the ability of hemoglobin to trap NO has also been used to "remove" NO from a system or to confirm its presence (in a negative control assay). Both oxy-hemoglobin ((Hb)Fe^{2+}–O$_2$; Eq. 4) and deoxy-hemoglobin ((Hb)Fe^{2+}; Eq. 5) react with NO (*reviewed in ref.* 8):

$$(Hb)Fe^{2+}\ O_2 + {}^{"}NO \rightarrow (Hb)Fe^{3+} + NO_3^{-} \qquad (4)$$

$$(Hb)Fe^{2+} + {}^{"}NO \rightarrow (Hb)Fe^{2+}\ NO \qquad (5)$$

Met-hemoglobin ((Hb)Fe^{3+}; Eq. 4), with its oxidized iron atom (Fe^{3+}, d^5 configuration), can be easily followed by EPR spectroscopy, with high sensitivity at low (liquid helium) temperatures ($\approx$100 nM, although the use of liquid helium increases the assays cost) [34]. However, there is no NO trapping and the formation of methemoglobin is not specific for NO, as peroxynitrite and other oxidants can also oxidize hemoglobin. Alternatively, one can follow the formation of the stable hemoglobin–NO complex ((Hb)Fe^{2+}–NO, $K_d \approx 10^{-12}$ to 10^{-10} M [8]; Eq. 5), at 77 K (liquid nitrogen temperature), without compromising the sensitivity ($\approx$200 nM) [34]. The assay has to be carried out under anaerobic conditions (to maintain the deoxygenation of hemoglobin) and the EPR signal is more complex in this case, but can be deconvoluted (by simulation) in three main species, penta- and hexacoordinated α-Hb-NO and hexacoordinated β-Hb-NO [34–38]. Furthermore, since NO is effectively trapped by hemoglobin, experiments with isotopically labeled nitrogen (^{15}N) can be carried out to undoubtedly identify the NO source (this is due to the fact that naturally abundant isotope of nitrogen (^{14}N) and ^{15}N have different nuclear magnetic moments (1 and 1/2) and thus originate different splittings).

The last type of spin-traps here described, and the one whose utilization is detailed discussed, makes use of the high affinity of NO to bind to ferrous chelates to detect it: iron- *bis* -dithiocarbamates

(Fe-(TC)$_2$) [39–45]. A dithiocarbamate (Fig. 2b) is a derivate of the carbamic acid (Fig. 2a), in which both oxygen atoms were replaced by sulfur—key to coordinate the iron atom—and whose amine function can be differently substituted (different R$_1$ and R$_2$, Fig. 2b) to produce hydrophobic and hydrophilic spin-traps (e.g., diethyldithiocarbamate (DETC) and *N*-methyl-D-glucamine dithiocarbamate (MGD), respectively; Fig. 2b). Biding of NO to a ferrous-*bis*-dithiocarbamate (Fe^{2+}-(TC)$_2$) results in formation of a mononitrosyl-iron complex (MNIC) (Fig. 2c; Eq. 6), that gives rise to a simple EPR signal that can be easily detected and analyzed at room temperature. The formation of MNIC is very fast ($k \approx 10^6$–10^8 M^{-1}s^{-1} [46–51]) and its stability is high (NO does not dissociate) as long as a high ratio of dithiocarbamate:iron is used ($\geq$10) [51].

$$Fe^{2+} \ (TC)_2 + {}^{\cdot}NO \rightarrow Fe^{2+} \ (TC)_2 \ NO(\{FeNO\}^7, paramagnetic) \qquad (6)$$

(The charge distribution in metal-NO adducts is frequently ambiguous; as such, the Enemark and Feltham notation was adopted,

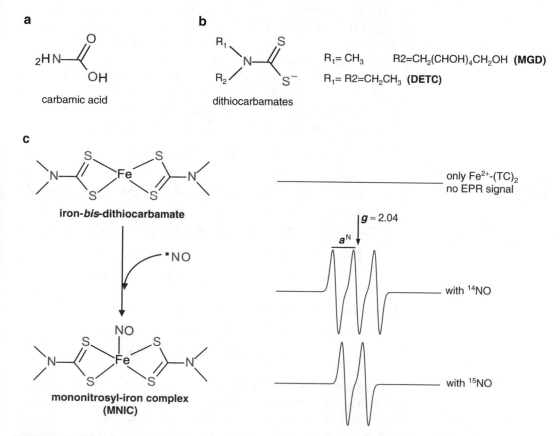

Fig. 2 Iron-dithiocarbamates. (**a**) Structure of the carbamic acid. (**b**) General structure of dithiocarbamates; amine substitutions (R$_1$ and R$_2$) of MGD and DETC are shown. (**c**) Structures of Fe-(TC)$_2$ and Fe-(TC)$_2$-NO and the respective characteristic EPR spectra Fe-(TC)$_2$ do not give rise to any EPR signal, while Fe-(TC)$_2$-^{14}NO and Fe-(TC)$_2$-^{15}NO yield a triplet and doublet signal, respectively)

{FeNO}n, where the superscript "n" indicates the number of metal d electrons plus the number of NO π-antibonding electrons [52].)

The MNIC EPR signal, with a g value of ≈ 2.04, is characterized by a 1:1:1 triplet pattern produced by the bound nitrogen atom of *natural* NO ($I(^{14}N) = 1$), with a nitrogen hyperfine splitting constant, a^N, of 1.2–1.3 mT. In the presence of isotopically labeled ^{15}NO, the resulting MNIC will give rise to the doublet signal characteristic of a nucleus with a I of 1/2, thus, allowing the clear identification of the NO source (e.g., study of NO generation in the presence of NO synthase and L-(^{15}N)-arginine).

The Fe-$(TC)_2$ have been widely used in in vitro studies, with purified and partially purified components [53–61], in vivo and in situ studies, with living animals, tissues, and cells [39–45, 62–70], using hydrophobic (e.g., DETC) and hydrophilic (e.g., MGD) spin-traps to target hydrophilic and lipophylic systems present in "aqueous compartments" and in membranes, respectively.

The Fe-$(TC)_2$ and the respective MNIC are, however, redox active species and care must be taken for possible artifacts. In aqueous solution, the Fe^{2+}-$(TC)_2$ are easily oxidized by dissolved dioxygen ($k \approx 10^5$–10^6 $M^{-1}s^{-1}$ [71, 72]) to the respective ferric complexes (Fe^{3+}-$(TC)_2$), thus, decreasing the concentration of ferrous complex available to detect NO. In addition, and more important, the Fe^{3+}-$(TC)_2$ formed rapidly consumes the NO (Eqs. 7–10; $k \approx 10^8$ $M^{-1}s^{-1}$ for Eq. 7 [46, 49–51]), but to yield an EPR signal whose intensity is less than 50 % of the one that would be originated by the ferrous complex (Eq. 11 vs. Eq. 6) [46, 51, 73–75]:

$$Fe^{3+} - (TC)_2 + \ ^{\cdot}NO \rightarrow Fe^{3+} - (TC)_2 - NO(\{FeNO\}^6, \text{diamagnetic}) \tag{7}$$

$$Fe^{3+} - (TC)_2 - NO \rightleftharpoons Fe^{2+} - (TC)_2 - NO^+ (\{FeNO\}^6, \text{diamagnetic}) \tag{(8)}$$

$$Fe^{2+} - (TC)_2 - NO^+ \underline{\quad} (H_2O \text{ or } HO^-) \rightarrow Fe^{2+} - (TC)_2 + NO_2^- \tag{9}$$

$$Fe^{3+} - (TC)_2 - NO + Fe^{2+} - (TC)_2 \rightarrow Fe^{3+} - (TC)_2 + Fe^{2+} - (TC)_2 - NO(\{FeNO\}^7, \text{paramagnetic}) \tag{10}$$

Global reaction (Eqs. $7 + 8 + 9 + 10$):

$$Fe^{3+} - (TC)_2 + 2\ ^{\cdot}NO \rightarrow 1Fe^{2+} \ (TC)_2 - NO(\{FeNO\}^7, \text{paramagnetic}) + NO_2^- \tag{11}$$

In this context, it should be here emphasized that Fe-$(TC)_2$ behaves differently in aqueous and organic solutions (e.g., the hydrophobic DETC–MNIC is not easily oxidized by oxygen and the diamagnetic complex is rapidly converted into the paramagnetic one in a hydrophobic medium [76]).

Fortunately, the oxidation of Fe^{2+}-$(TC)_2$ is easily detected by UV–visible absorption (even at naked eye), as the colorless ferrous complex solution is converted into an orange-brown solution (details in the Subheading 3) [51, 75]. The potential formation of Fe^{3+}-$(TC)_2$ can also be monitored by EPR at 77 K, following the $g \approx 4.3$ signal characteristic of high-spin Fe^{3+} ($S = 5/2$) [51, 74].

In a similar way, oxidation of Fe^{2+}-$(TC)_2$-NO and Fe^{2+}-$(TC)_2$ by oxidants present in the sample to be analyzed would falsely lead to a decreased EPR signal and, consequently, to an underestimation of the NO present. Yet, the *other side of the coin* is that reducers present in the sample (like ascorbate [51, 72, 77], GSH and L-cysteine [72, 78]) can promote the reduction of Fe^{3+}-$(TC)_2$-NO to the paramagnetic species. Hence, the ratio between reduced and oxidized Fe-$(TC)_2$ and Fe-$(TC)_2$-NO (particularly in cells, tissues and living organisms) would depend on the presence of endogenous and exogenous oxidants and reducers and on the presence of endogenous or exogenous superoxide dismutase and catalase (discussed below).

To sum up, the redox properties of the Fe-$(TC)_2$ and respective MNIC could greatly complicate the quantitative analysis of NO concentration. Accordingly, when possible and reasonable, care must be taken to ensure anaerobic and reducing conditions that preserve the Fe-$(TC)_2$ and Fe-$(TC)_2$-NO complexes in the ferrous state. This would not be possible, or even advisable, in many biological systems or in in vivo and in situ assays. Nevertheless, and most important, a qualitative or comparative analysis could still be made, since Fe^{3+}-$(TC)_2$ would also lead to the formation of measurable paramagnetic MNIC.

Other concern relates to the specificity of the Fe-$(TC)_2$ complexes for NO. There is some controversy regarding their ability to originate paramagnetic complexes with the nitrosonium cation (NO^+) donor sodium nitroprusside and with the nitroxyl anion (NO^-) donor Angeli's salt. The complex resultant from the reaction of Fe^{2+}-$(TC)_2$ with nitroxyl would be EPR silent ($\{FeNO\}^8$) [79], and, while Xia and Zweier [54] in fact did not observed any EPR signal in the presence of Angeli's salt, Komarov et al. [80] did obtain a signal. However, as Xia et al. [81] suggested, the nitroxyl is probably reacting with the oxidized trap (to yield a $\{FeNO\}^7$), which, by intramolecular iron reduction, would result in the expected paramagnetic Fe^{2+}-$(TC)_2$-NO. In its turn, the paramagnetic complex responsible for the nitrosonium signal in the presence of Fe^{2+}-$(TC)_2$ (an initial $\{FeNO\}^6$ complex) [80] could be formed in a similar way to the one suggested for the signal obtained with NO and Fe^{3+}-$(TC)_2$ (also an initial $\{FeNO\}^6$ complex that eventually yields an $\{FeNO\}^7$ species (Eqs. 7–10)). In addition, also S-nitrosoglutathione (GSNO) was shown to elicit the formation of paramagnetic MNIC [78]. In contrast, there is no debate concerning the inability of Fe-$(TC)_2$ to yield EPR signals with nitrite [56, 58–61, 80]. Furthermore, although nitrite was described to be able to oxidize Fe^{2+}-$(TC)_2$, the reaction is too slow to be of any practical relevance ($\approx 5 \ M^{-2}s^{-1}$) [75].

Other issues can be raised regarding reactive oxygen species (superoxide radical anion, hydrogen peroxide, hydroxyl radical). Although they can be "normal" products of the system under study, reactive oxygen species can also arise as "by-products" of the dioxygen oxidation of Fe-$(TC)_2$ and, thus, introduce "extra" oxidative damages to the biomolecules ("extra" oxidative stress) or cause artificial NO generation from nitrogenous compounds [71, 78]. Moreover, hydrogen peroxide [72], superoxide and peroxynitrite [48, 77] are able to convert paramagnetic MNIC into EPR silent complexes. Although the reaction of NO with superoxide (to yield peroxynitrite) is considerably faster ($k \approx 7 \times 10^9 M^{-1} s^{-1}$ [12, 13]) than its reaction with Fe-$(TC)_2$, the presence of millimolar Fe-$(TC)_2$ (comparatively to the low (micromolar at most) superoxide concentration normally found) would guarantee the formation of (at least some) Fe-$(TC)_2$-NO. However, the formed Fe^{2+}-$(TC)_2$-NO, present at low concentration, cannot escape to the "attack" of superoxide ($k \approx 3 \times 10^7 M^{-1} s^{-1}$ [77]). In this context, the presence of exogenous and endogenous superoxide dismutase and catalase has to be carefully thought when quantitative analyses of NO concentration are made.

Finally, it should not be overlooked that the Fe-$(TC)_2$ and Fe-$(TC)_2$-NO complexes can also interfere directly with the system under study by inhibiting or inactivating enzymes, receptors, and many other biomolecules. Appropriate controls should be done to guarantee that, e.g., the absence (or weak) of EPR signal is not due to inhibition of the NO generating system; or the opposite, that an intense EPR signal is not the result of the inactivation of the NO target or scavenger. In this respect, particular attention has to be paid to in vivo and in situ studies (where the complexity of the system is very high) and to the use of different Fe-$(TC)_2$ spin-traps, since different substitutions (R_1 and R_2, Fig. 2b) could have dissimilar effects on the biomolecules and be differently accumulated subcellularly. The chelating activity of dithiocarbamates should never be disregarded. Dithiocarbamates are powerful chelators that can chelate free metal ions as well as metal ions present in biomolecules and, in this way, inactivate those biomolecules (e.g., the Cu, Zn-superoxide dismutase inhibition by DETC was attributed to the removal of the enzyme essential copper atom [82]). On the other hand, also the addition of iron can be detrimental (at least through the catalysis of radical species formation).

To conclude, the use each type of spin-trap has its advantages and disadvantages. In this chapter, particular attention was given to Fe-$(TC)_2$ that are the NO traps of choice for the authors. Certainly, the redox chemistry and nature of Fe-$(TC)_2$ and Fe-$(TC)_2$-NO can introduce artifacts and drawbacks in their utilization (Fig. 3). Nonetheless, the careful planning of the assays allows one to avoid the *problems* and use the Fe-$(TC)_2$ as a valuable approach to obtain accurate and reproducible results that cannot be easily acquired other means.

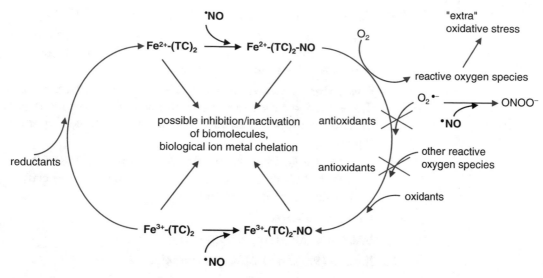

Fig. 3 Possible artifacts introduced by the redox chemistry and nature of Fe-(TC)$_2$ and Fe-(TC)$_2$-NO complexes

1.3 Spin-Trapping with Iron-Dithiocarbamates

The NO spin-trapping with iron-dithiocarbamates methodology is here exemplified using the ability of mammalian xanthine oxidase (XO) and bacterial aldehyde oxidoreductase (AOR) to reduce nitrite to generate NO [60, 61]. The objective is to demonstrate and quantify the NO formation. For that purpose, the protocol is very simple to perform: it involves mixing the enzyme, a reducing substrate and nitrite in the presence of the spin-trap, in a buffered system. The EPR spectrum is, then, acquired. The observation of a triplet signal with the characteristic parameters confirms the presence of NO. The quantification is made with a standard curve prepared with NO solutions of known concentration or by calibration with a nitroxide radical of known concentration. The NO formation over time can be followed through the increase of the signal intensity.

2   Materials

2.1 Preparation of Solutions and Materials for EPR Measurements

Preferably, use in all solutions water first distilled and then deionized (with $\leq$18 MΩ (25 °C)), to avoid adding unwanted redox active metal ions and other compounds that may interfere with the redox chemistry of the spin-trap and with your system (many enzymes are inhibited by heavy metal ions).

1. MGD (also named N-(dithiocarboxy)-N-methyl-D-glucamine), sodium salt, with a purity $\geq$98.0 % (for EPR spectroscopy).

2. Ferrous ammonium sulfate, with a high purity (99.997 %).

3. NO gas, $\approx$100 % or any percentage prepared by the gases supplier, e.g., 5 % NO/95 % He.

4. Gaseous argon or nitrogen with high purity (>99.999 %).

5. Proper tube to delivery the gases (*see* Fig. 4a, b), which should not be permeable to argon, nitrogen, NO, and oxygen and other gases from the atmosphere.

6. Flasks with a neck suitable to be closed by a rubber stopper and the respective rubber stoppers; aluminum crimp caps (to fix the rubber stoppers, if needed) and a hand-operated aluminum cap crimper.

7. Needles; we find the ones with 0.9×90 mm and 0.8×50 mm suitable to bubble the gas and to act as "escape," respectively (dimensions are diameter×length).

8. *Gastight* syringes.

9. Flat cell or 50 μL capillaries.

10. X-band (9.5 GHz) EPR spectrometer.

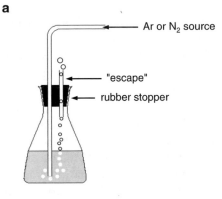

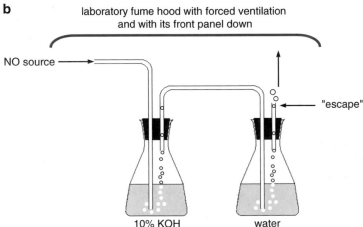

Fig. 4 Settings used to prepare deoxygenated solutions (**a**) and the stock NO saturated solution (**b**)

2.2 For the AOR and XO-Dependent NO Generation Assays

1. Buffer: 50 mM phosphate buffer, pH 6.0–7.8, 100 mM MES buffer for pH 5.5–6.0; 100 mM Tris–HCl buffer is also suitable for pH 7–9.

2. The enzymes used are purified, but partially purified samples or crude extracts can also be used.

3. Nitrite.

4. Reducing substrate, xanthine for the XO reaction and benzaldehyde for the AOR reaction (*see* **Note 1**).

3 Methods

As described in the "Introduction," if the system and study aims allow it, take great care to guarantee anaerobic (deoxygenated) conditions throughout all the procedure (*see* **Note 2**). In this way, the Fe-(MGD)$_2$ and Fe-(MGD)$_2$-NO complexes would be maintained in the ferrous state (*see* **Note 3**). The anaerobic conditions can be achieved by using flasks sealed with rubber stoppers and purging the solutions or water with an inert gas, nitrogen, or argon (*see* **Note 4**) (Fig. 4a). Also the MGD and iron powders have to be deoxygenated in sealed flaks before the addition of the solution in which they will be dissolved. To transfer the solutions/water from one flask to another or to the EPR flat cell, use *gastight* syringes (syringes whose plunger tip, often in PTFE, creates a leak-free seal, e.g., Hamilton® GASTIGHT® Syringes). During the transfer operations, be very careful to avoid contaminations by oxygen from the atmosphere. Ideally, the assay mixtures should be prepared inside an anaerobic chamber (*see* **Note 5**), as this allows one to manipulate the solutions and powders without taking the risk of contaminating them with the atmospheric oxygen.

3.1 Preparation of Fe-(MGD)$_2$

1. The two solids, MGD and ferrous ammonium sulfate, should be brought together and deoxygenated in a sealed flask; subsequently, deoxygenated buffer or water should be added to dissolve them.

2. Prepare the solution only immediately before use—do not use "old" solutions.

3. Weight the necessary masses to prepare a mixture of 20 mM MGD:2 mM iron, final concentrations.

4. A high MGD:iron ratio is important to ensure the stability of the future Fe^{2+}-(MGD)$_2$-NO complex (*see* **Note 6**).

5. However after a first trial, you may want to test a lower ratio to save MGD. In our assays, we use a ratio of 10 (5 mM MGD:0.5 mM iron) with the DEAE NO donor, while a ratio of 5 (10 mM MGD:2 mM iron, final concentrations) results very well with purified enzymes [60, 61].

The Fe^{2+}-$(MGD)_2$ solution should be light colorless and the potential formation of ferric complexes can be easily followed and quantified through the UV–visible spectrum due to their orange-brown coloration [maximum absorption at 340 nm ($\varepsilon = 20{,}000$ M^{-1} cm^{-1}) and shoulders at 385 nm ($\varepsilon = 15.000$ M^{-1} cm^{-1}) and 520 nm ($\varepsilon = 3.000$ M^{-1} cm^{-1}).

3.2 Preparation of Stock NO Saturated Solution

1. The saturated solution of NO is prepared by bubbling NO gas, first, in a potassium hydroxide 10 % solution, to remove other nitrogen oxides present in the NO gas, and, then, in water, as shown in Fig. 4b (*see* **Notes 7** and **8**).

2. It is crucial that the potassium hydroxide solution is prepared in deoxygenated, deionized water, and the "final" water (where NO gas will be dissolved) should also be deoxygenated and deionized (follow the indications provided in the beginning of this section on how to deoxygenate solutions).

3. The NO purging should be performed in an ice-water bath, because lower temperatures increase the NO solubility in solution (*see* **Note 9**). The NO bubbling time is dependent on the NO gas concentration: for a pure gas ($\approx$100 % NO), 10 min would be sufficient, but, for a 5 % gas (5 % NO/95 % He), purge for at least 30 min.

4. After that time, close the NO source and, immediately, remove the two needles from the water flask (*see* **Note 10**).

5. The concentration of NO dissolved in water is dependent on the percentage of NO gas used and on the temperature: for $\approx$100 % NO, it is $\approx$3.3 and $\approx$1.91 mM, at 0 and 20 °C, respectively, and for 5 % NO, it is $\approx$170 and $\approx$100 µM, at 0 and 20 °C, respectively. Dilutions can be, then, prepared from the stock NO saturated solution (*see* **Note 11**).

3.3 Preparation of Reaction Mixtures

1. Deoxygenate all solutions immediately before being used.

2. Prepare your reactions mixtures starting with the buffer and spin-trap; the NO generating/consuming system should be the last to be added (this is the moment that establishes the "time zero" of the reaction and the spin-trap must already be present to trap the NO).

3. For the AOR- and XO-catalyzed nitrite-dependent NO generation, this means that the additions should be made by the following order: buffer (and water if necessary), spin-trap, enzyme, reducing substrate and at least nitrite.

4. Usually, the measurements are carried out in 275–300 µL flat cells (*see* **Note 12**), but 50 µL capillaries can also be used to save sample, if it has a sufficiently high concentration of paramagnetic species to give rise to a measurable signal with the smaller sample amount.

5. If an anaerobic chamber is used to prepare the reactions mixtures, filling the flat cell or the capillary should be an easy task, using a simple glass pipette, or by capillarity, respectively (both ends of the capillary should then be sealed).

6. But, if an anaerobic chamber is not available, it would be very difficult to work with capillaries and maintain the anaerobic conditions.

7. Regarding the flat cell, this must be first sealed with a rubber stopper and made anaerobic; after that, the sample can be transferred with a *gastight* syringe.

**3.4 EPR
Measurements**

1. The EPR signal of the Fe^{2+}-$(MGD)_2$-NO complex has a g value of ≈ 2.04, an a^N of 1.2–1.3 mT and an 1:1:1 triplet or 1:1 doublet pattern, depending if it is originated from ^{14}NO or isotopically labeled.

2. In an X-band EPR spectrophotometer, with a frequency of 9.7 GHz, the signal will be centered around 340 mT (Eq. 2b) and a window width of 8 mT would be ample to display the signal.

3. A modulation amplitude of 0.1–0.3 mT is used and the microwave power needed would depend on the concentration of paramagnetic complexes present; usually a value between 10 and 100 mW is employed (*see* **Note 13**).

4. The observation of the characteristic signal confirms the presence of NO in the sample.

5. For the AOR and XO assays, the use of *normal* nitrite (that is, with naturally abundant isotope of nitrogen (^{14}N)) and of isotopically labeled $^{15}NO_2^-$, results in the expected triplet and doublet signals, respectively. These two results demonstrate that is nitrite the source of NO.

6. To quantify the concentration of NO (or more precisely of Fe^{2+}-$(MGD)_2$-NO), a standard curve can be prepared using the spin-trap and solutions equilibrated with different NO concentrations prepared from the stock NO saturated solution.

7. It is crucial, that the standard curve is obtained under the same conditions as the assays, same buffer composition (buffer species and other compounds present, concentration, pH, and ionic strength), temperature, and spin-trap concentration.

8. The EPR signal is exactly the same in both cases, the signal intensity of the assay and standard can be measured by the height of one of its three lines, or by the double integral (*see* **Note 14**)—what is easier for the researcher.

9. Quantification can also be performed by comparison with the signal of 2,2,6,6-tetramethylpiperidinyl-*N*-oxyl (TEMPO). TEMPO is a nitroxide radical that gives rise to a triplet signal,

with an a^N of ≈ 1.75 mT, and the double integral of its signal provides a relation area—concentration that can be subsequently used to convert the areas of the signal of the assays into concentration values (*see* **Note 15**).

10. In general, Fe^{2+}-$(MGD)_2$-NO as low as 10 nM can be detected, with the advantage that also kinetics of NO increase can be easily followed in one unique sample.

11. Because of the high stability of the Fe^{2+}-$(MGD)_2$-NO complex, NO decreases over time could not be measured in the only one sample.

12. Yet, different samples can be prepared to be read at different times.

13. For the AOR and XO assays, the concentration of NO formed is dependent on the reducing substrate and nitrite concentration, on the pH and also on the enzyme concentration/time of reaction.

14. Different proportions of reducing substrate/nitrite should be assayed to find the ones that give rise to better results at each pH value.

3.5 Controls

As in other methodologies, it is essential to carry out the appropriate controls to guarantee that the conclusions drawn are correctly supported.

The first obligate controls involve the study of the potential Fe^{2+}-$(MGD)_2$-NO formation in the absence of each individual component of the system. For the AOR- and XO-catalyzed nitrite-dependent NO generation, the Fe^{2+}-$(MGD)_2$-NO formation should be studied in the absence of nitrite (presence of all the other components), absence of enzyme (presence of all the other components), and absence of reducing substrate (presence of all the other components). No observation of EPR signal in the absence of any of the three components demonstrates that it is the enzyme (AOR or XO) that is responsible for the NO formation. The observation of EPR signal in the presence of reducing substrates of different nature further confirms the XO ability to form NO (*see* **Note 16**). To discriminate if it is an enzymatic reaction, or, on the contrary, is just a nonspecific reaction carried out by the protein moiety, test the inactivated or denatured enzyme.

If no EPR signal is observed in a situation where it is expected the NO formation, investigate the interference of Fe^{2+}-$(MGD)_2$ and Fe^{2+}-$(MGD)_2$-NO with the system, for example, the potential enzyme inhibition.

Another control that can be helpful in case of no EPR signal formation is to test if the *medium/conditions* of the system interfere with the Fe^{2+}-$(MGD)_2$-NO formation. For this, test the ability of Fe^{2+}-$(MGD)_2$ to generate the paramagnetic species with the stock NO saturated solution under the same *medium/conditions* used in the system under study. If, hypothetically, the buffer hinders the formation of the Fe^{2+}-$(MGD)_2$-NO complex, then no

EPR signal will be formed either with the stock NO solution or with your system.

Pay special attention to the controls, if you use partially purified enzymes or crude extracts, as the number of problems that can arise are proportional to the complexity of the system under study.

4   Notes

1. The reducing substrate reacts with the enzyme; the reduced enzyme, then, reduces nitrite to NO, closing the catalytic cycle. Other reducing substrates of each enzyme can be equally employed.

2. However, even if the Fe^{2+}-$(MGD)_2$ solution is mostly oxidized to the ferric state, the formation of measurable paramagnetic Fe^{2+}-$(MGD)_2$-NO may not be hindered (via Eq. 11 and through reducing compounds present in the sample that can reduce the ferric complexes back to Fe^{2+}-$(MGD)_2$ and Fe^{2+}-$(MGD)_2$-NO). The possible presence of antioxidants, as well as, of oxidants must be considered.

3. Note that the preparation of Fe-$(MGD)_2$ under aerobic conditions leads to its rapid and uncontrolled oxidation to Fe^{3+}-$(MGD)_2$ prior to its utilization. Moreover, after addition to the system, the concomitantly reduced oxygen would scavenge the NO and increase the oxidative stress induced in the system. These should affect (diminish) the reproducibility and accuracy of the data to be obtained. Therefore, even if the assays are to be performed under aerobic conditions, prepare the spin-trap under anaerobic conditions and tightly control the timing when it becomes in contact with oxygen.

4. To deoxygenate a solution/water (or a powder), the dissolved oxygen must be removed, being replaced by an inert gas, nitrogen, or argon. To do this, bubble the inert gas into the solution/water (it can be done under stirring to facilitate the gases exchange). For oxygen to be released (and do not build in pressure), a pressure relief must be provided during purging (an "escape"), which is usually achieved by the insertion of a second needle that must not touch the liquid, as illustrated in Fig. 4a. The bubbling time should be proportional to the volume of liquid to be deoxygenated; for 10–20 mL, 30 min would be sufficient. The efficiency of deoxygenation can be greatly improved if purging is intercalated by an equal time period of vacuum (usually three cycles of purging/vacuum are sufficient). Regarding the choice of the inert gas: argon is more expensive than nitrogen, but, because it is heavier, its "deoxygenation ability" is more efficient. Obviously, this procedure also removes carbon dioxide and other atmospheric gases that are dissolved in the solutions/water.

5. Please refer to the instructions of the respective anaerobic chamber (also known as glove box) before attempting to use it. Although all solutions and powders must be made anaerobic before entering the anaerobic chamber (that is, that part cannot be circumvented), the use of an anaerobic chamber avoids the problems of oxygen contaminations during the transfer and mixing procedures.

6. In in vivo or in situ assays, guarantee that MGD and iron are added in sufficient concentration to facilitate the formation of a high MGD:iron ration inside the cells; also consider the solubility of the trap chosen (MGD and DETC accumulate in aqueous and lipidic compartments).

7. The NO gas handling must be performed only in a laboratory fume hood with forced ventilation and with its front panel down. The personnel handling the gas must be thoroughly familiar with the Material Safety Data Sheet and with the proper handling procedures.

8. With the setting shown in Fig. 4b, be particularly cautious, because the gas "escape" is present only in the water flask. You must confirm that NO is bubbling through the KOH solution to the "final" water flask, not only for safety reasons, but also to guarantee that the water is in fact being saturated with NO.

9. At room temperature, the NO solubility would decrease, but, in this way, when you use the solution at room temperature you will be sure that the solution is truly saturated with NO.

10. If the two needles are not removed immediately, the oxygen from the atmosphere will get in into the solution through the "escape" needle and the NO concentration will be decreased (NO reacts with oxygen to yield nitrogen dioxide radical and other products).

11. When working outside an anaerobic chamber, do not forget to be very careful during the dilutions to avoid contamination with the oxygen from the atmosphere.

12. The common 4 mm (internal diameter) tube cannot be used in X-band EPR spectroscopy at room temperature of aqueous solutions, because the water present in the sample hinders the measurements: the electric dipole moment of the water molecule interacts with the electric field in the resonator and a large amount of the microwave energy is absorbed (lost) by the water. To circumvent the problem created by liquids with dielectric loss (in X-band EPR cavities) one can use flat cells, which confine the sample to a 0.3 mm thick space, in a volume typically of 275–300 µL. The flat cell has to be placed in the rectangular cavity exactly perpendicularly to the external magnets, parallel to the electric field, in a nodal plane where the electric field intensity is minimal and the magnetic field is maxi-

mal; a few degrees of misalignment of the cell inside the cavity results in a sharp increase in the dissipated (lost) energy, with the concomitant decrease in the energy stored to drive the absorption by EPR. Alternatively, a 50 μL capillary, with an internal diameter of 0.8 mm, can solve the problem if the sample has a sufficiently high concentration of paramagnetic species to give rise to a measurable signal with the small sample amount.

13. Confirm that your signal is not saturated. For a nonsaturated signal, the intensity is directly proportional to the square root of the power. The intensity can be determined by measuring the height of one of the signal three lines, or by the double integral (*see also* **Note 14**).

14. Note that it is necessary to calculate the double integral, because the EPR spectrum is recorded as the first derivative of absorption (double integral = area under the absorption line).

15. The direct comparison of EPR spectra is possible, because the intensity of an EPR signal is independent of the species that has originated it (it is not necessary the previous knowledge of an extinction coefficient). The use of TEMPO or NO saturated solution may depend on their availability. The usage of a NO saturated solution has the advantage of providing hints of possible problems arising from the assay system/mixture, if it is possible to compare the results obtained in different media.

16. It is known that XO displays catalytic activity (oxygen-reducing activity) with reducing substrates of different nature (e.g., purines and aldehydes), being classified as a promiscuous enzyme.

Acknowledgments

This work was supported by the Unidade de Ciências Biomoleculares Aplicadas-UCIBIO which is financed by national funds from FCT/MEC (UID/Multi/04378/2013) and co-financed by the ERDF under the PT2020 Partnership Agreement (POCI-01-0145-FEDER-007728.

References

1. Moncada S, Palmer RMJ, Higgs EA (1991) Nitric oxide: physiology, pathophysiology, and pharmacology. Pharmacol Rev 43:109–142
2. Gusarov I, Nudler E (2005) NO-mediated cytoprotection: instant adaptation to oxidative stress in bacteria. Proc Natl Acad Sci U S A 102:13855–13860
3. Neill S, Bright J, Desikan R, Hancock J, Harrison J, Wilson I (2008) Nitric oxide evolution and perception. J Exp Bot 59:25–35
4. Wilson ID, Neill SJ, Hancock JT (2008) Nitric oxide synthesis and signalling in plants. Plant Cell Environ 31:622–631
5. Gupta KJ, Kaiser WM (2010) Production and scavenging of nitric oxide by barley root mitochondria. Plant Cell Physiol 51:576–584
6. Moreau M, Lindermayr C, Durner J, Klessig DF (2010) NO synthesis and signaling in plants -where do we stand? Physiol Plant 138:372–383

7. Toledo JC, Augusto O (2012) Connecting the chemical and biological properties of nitric oxide. Chem Res Toxicol 25:975–989

8. Maia L, Moura JJG (2014) How biology handles nitrite. Chem Rev 114:5273–5357

9. Dalton LR (ed) (1985) EPR and advanced EPR studies of biological systems. CRC Press, Boca Raton

10. Palmer G (1985) The electron paramagnetic resonance of metalloproteins. Biochem Soc Trans 13:548–560

11. Weil JA, Bolton JR, Wertz JE (1994) Electron paramagnetic resonance: elementary theory and practical applications. Wiley, New York

12. Beckman JS, Koppenol WH (1996) Nitric oxide, superoxide and peroxynitrite: the good, the bad and the ugly. Am J Physiol 271:C1424–C1437

13. Koppenol WH (1998) The basic chemistry of nitrogen monoxide and peroxynitrite. Free Radic Biol Med 25:385–391

14. Wink DA, Mitchell JB (1998) Chemical biology of nitric oxide: insights into regulatory, cytotoxic, and cytoprotective mechanisms of nitric oxide. Free Radic Biol Med 25:434–456

15. Feelisch M, Rassaf T, Mnaimneh S, Singh N, Bryan NS, Jourd'Heuil D, Kelm M (2002) Concomitant S-, N-, and heme-nitros(yl) ation in biological tissues and fluids: implications for the fate of NO in vivo. FASEB J 16:1775–1785

16. Taha ZH (2003) Nitric oxide measurements in biological samples. Talanta 61:3–10

17. Bryan NS, Rassaf T, Maloney RE, Rodriguez CM, Saijo F, Rodriguez JR, Feelisch M (2004) Cellular targets and mechanisms of nitros(yl) ation: an insight into their nature and kinetics in vivo. Proc Natl Acad Sci U S A 101:4308–4313

18. Perkins MJ (1980) Spin trapping. Adv Phys Org Chem 17:1–64

19. Janzen EG, Haire DL (1990) Two decades of spin trapping. Adv Free Rad Chem 1:253–295

20. Davies MJ, Timmins GS, Clark RJH, Hester RE (1996) EPR spectroscopy of biologically relevant free radicals in cellular, ex vivo, and in vivo systems. In: Clark RJH, Hester RE (eds) Biomedical applications of spectroscopy. John Wiley & Sons, New York

21. Berliner LJ, Khramtsov V, Fujii H, Clanton TL (2001) Unique in vivo applications of spin traps. Free Radic Biol Med 30:489–499

22. Villamena FA, Zweier JL (2004) Detection of reactive oxygen and nitrogen species by EPR spin trapping. Antioxid Redox Signal 6:619–629

23. Davies MJ (1993) Detection and identification of macromolecule-derived radicals by EPR spin trapping. Res Chem Intermed 19:669–679

24. Clement JL, Gilbert BC, Rockenbauer A, Tordo P (2001) Radical damage to proteins studied by EPR spin-trapping techniques. J Chem Soc Perkin Trans 2:1463–1470

25. Davies MJ, Hawkins CL (2004) EPR spin trapping of protein radicals. Free Radic Biol Med 36:1072–1086

26. Akaike T, Yoshida M, Miyamoto Y, Sato K, Kohno M, Sasamoto K, Miyazaki K, Ueda S, Maeda H (1993) Antagonistic action of imidazolineoxyl N-oxides against endothelium-derived relaxing factor/.bul.NO (nitric oxide) through a radical reaction. Biochemistry 32:827–832

27. Joseph J, Kalyanaraman B, Hyde JS (1993) Trapping of nitric oxide by nitronyl nitroxides: an electron spin resonance investigation. Biochem Biophys Res Commun 192:926–934

28. Hogg N, Singh RJ, Joseph J, Neese F, Kalyanaraman B (1995) Reactions of nitric oxide with nitronyl nitroxides and oxygen: prediction of nitrite and nitrate formation by kinetic simulation. Free Radic Res 22:47–56

29. Grätzel M, Taniguchi S, Henglein Ber A (1970) Pulsradiolytische Untersuchung der NO-Oxydation und des Gleichgewichts $N_2O_3 \rightarrow NO + NO_2$ in wäßriger Lösung. Ber Bunsen-Ges Phys Chem 74:488–492

30. Treinin A, Hayon E (1970) Absorption spectra and reaction kinetics of NO_2, N_2O_3, and N_2O_4 in aqueous solution. J Am Chem Soc 92:5821–5828

31. Singh RJ, Hogg N, Joseph J, Konorev E, Kalyanaraman B (1999) The peroxynitrite generator, SIN-1, becomes a nitric oxide donor in the presence of electron acceptors. Arch Biochem Biophys 361:331–339

32. Zhang Y, Hogg N (2004) Formation and stability of S-nitrosothiols in RAW 2647 cells. Am J Physiol Lung Cell Mol Physiol 287:L467–L474

33. Zhang Y, Hogg N (2002) Mixing artifacts from the bolus addition of nitric oxide to oxymyoglobin: implications for s-nitrosothiol formation. Free Radic Biol Med 32:1212–1219

34. Piknova B, Gladwin MT, Schechter AN, Hogg N (2005) Electron paramagnetic resonance analysis of nitrosylhemoglobin in humans during NO inhalation. J Biol Chem 280:40583–40588

35. Hille R, Olson JS, Palmer G (1979) Spectral transitions of nitrosyl hemes during ligand binding to hemoglobin. J Biol Chem 254:12110–12120

36. Louro SR, Ribeiro PC, Bemski G (1981) EPR spectral changes of nitrosyl hemes and their relation to the hemoglobin T-R transition. Biochim Biophys Acta 670:56–63

37. Kozlov AV, Staniek K, Nohl H (1999) Nitrite reductase activity is a novel function of mammalian mitochondria. FEBS Lett 454:127–130

38. Huang KT, Keszler A, Patel N, Patel RP, Gladwin MT, Kim-Shapiro DB, Hogg N (2005) The reaction between nitrite and deoxyhemoglobin - reassessment of reaction kinetics and stoichiometry. J Biol Chem 280: 31126–31131

39. Vanin AE, Mordvintcev EI, Kleschyov AL (1984) Biotransformation of sodium nitroprusside into dinitrosyl iron complexes in tissue of ascites tumors of mice. Stud Biophys 102:135–143

40. Varich VJ, Vanin AE, Ovsyannikova LM (1987) Discovery of endogenous nitric oxide in the mouse liver by electron paramagnetic resonance. Biofizika 32:1064–1065

41. Mordvintcev P, Mülsch A, Busse R, Vanin A (1991) On-line detection of nitric oxide formation in liquid aqueous phase by electron paramagnetic resonance spectroscopy. Anal Biochem 199:142–146

42. Kubrina LN, Caldwell WS, Mordvintcev EI, Malenkova IV, Vanin AF (1992) EPR evidence for nitric oxide production from guanidino nitrogen of L-arginine in animal tissues in vivo. Biochim Biophys Acta 1099:233–237

43. Komarov A, Mattson D, Jones MM, Singh PK, Lai CS (1993) In vivo spin trapping of nitric oxide in mice. Biochem Biophys Res Commun 195:1191–1198

44. Komarov AM, Lai CS (1995) Detection of nitric oxide production in mice by spin-trapping electron paramagnetic resonance spectroscopy. Biochim Biophys Acta 1272:29–36

45. Vanin AF, Huisman A, Van Faassen E (2002) Iron dithiocarbamate as spin trap for nitric oxide detection: pitfalls and successes. Methods Enzymol 359:27–42

46. Fujii S, Yoshimura T, Kamada H (1996) Nitric oxide trapping efficiencies of water-soluble iron(III) complexes with dithiocarbamate derivatives. Chem Lett 1996:785–786

47. Paschenko SV, Khramtsov VV, Scatchkov MP, Plyusnin VF, Bassenge E (1996) EPR and laser flash photolis studies of the reaction of nitric oxide with water soluble NO trap Fe(II)-proline-dithiocarbamate complex. Biochem Biophys Res Commun 225:577–584

48. Pou S, Tsai P, Porasuphatana S, Halpern HJ, Chandramouli GV, Barth ED, Rosen GM (1999) Spin trapping of nitric oxide by ferrochelates: kinetic and in vivo pharmacokinetic studies. Biochim Biophys Acta 1427:216–226

49. Fujii S, Kobayashi K, Tagawa S, Yoshimura T (2000) Reaction of nitric oxide with the iron(III) complex of N-(dithiocarboxy)sarcosine: a new type of reductive nitrosylation involving iron(IV) as an intermediate. J Chem Soc Dalton Trans:3310–3315

50. Fujii S, Yoshimura T (2000) A new trend in iron–dithiocarbamate complexes: as an endogenous NO trapping agent. Coord Chem Rev 198:89–99

51. Vanin AF, Liu XP, Samouilov A, Stukan RA, Zweier JL (2000) Redox properties of iron–dithiocarbamates and their nitrosyl derivatives: implications for their use as traps of nitric oxide in biological systems. Biochim Biophys Acta 1474:365–377

52. Enemark JH, Feltham RD (1974) Principles of structure, bonding, and reactivity metal nitrosyl complexes. Coord Chem Rev 13:339–406

53. Miilsch A, Vanin A, Mordvintcev R, Hauschildt S, Busse R (1992) NO accounts completely for the oxygenated nitrogen species generated by enzymic L-arginine oxygenation. Biochem J 288:597–603

54. Xia Y, Zweier JL (1997) Direct measurement of nitric oxide generation from nitric oxide synthase. Proc Natl Acad Sci U S A 94: 12705–12710

55. Yoneyema H, Kosaka H, Ohnishi T, Kawazoe T, Mizoguchi K, Ichikawa Y (1999) Reaction of neuronal nitric oxide synthase with the nitric oxide spin-trapping agent, iron complexed with N-dithiocarboxysarcosine. Eur J Biochem 266:771–777

56. Li H, Samouilov A, Liu X, Zweier JL (2001) Characterization of the magnitude and kinetics of xanthine oxidase catalyzed nitrite reduction: evaluation of its role in nitric oxide generation in anoxic tissues. J Biol Chem 276:24482–24489

57. Huisman A, Vos I, van Faassen EE, Joles JA, Grone HJ, Martasek P, Zonneveld AJ, Vanin AF, Rabelink TJ (2002) Anti-inflammatory effects of tetrahydrobiopterin on early rejection in renal allografts: modulation of inducible nitric oxide synthase. FASEB J 16:1135–1137

58. Li H, Cui H, Kundu TK, Alzawahra W, Zweier JL (2008) Nitric oxide production from nitrite occurs primarily in tissues not in the blood: Critical role of xanthine oxidase and aldehyde oxidase. J Biol Chem 283:17855–17863

59. Li H, Kundu TK, Zweier JL (2009) Characterization of the magnitude and mechanism of aldehyde oxidase-mediated nitric oxide production from nitrite. J Biol Chem 284: 33850–33858

60. Maia L, Moura JJG (2011) Nitrite reduction by xanthine oxidase family enzymes: a new class of nitrite reductases. J Biol Inorg Chem 16:443–460

61. Maia L, Moura JJG (2014) Nitrite reductase activity of rat and human xanthine oxidase, xanthine dehydrogenase, and aldehyde oxidase: evaluation of their contribution to NO formation in vivo. Biochemistry. doi:10.1021/bi500987w

62. Tominaga T, Sato S, Ohnishi T, Ohnishi ST (1993) Potentiation of nitric oxide formation following with bilateral carotid occlusion and

local cerebral ischemia in the rat: in vivo detection of the nitric oxide radical by electron paramagnetic resonance spin trapping. Brain Res 614:342–346

63. Lai CS, Komarov AM (1994) Spin trapping of nitric oxide produced in vivo in septic-shock mice. FEBS Lett 345:120–124

64. Zweier JL, Wang P, Kuppusamy P (1995) Direct measurement of nitric oxide generation in the ischemic heart using electron paramagnetic resonance spectroscopy. J Biol Chem 270:304–307

65. Zweier JL, Wang P, Samouilov A, Kuppusamy P (1995) Enzyme-independent formation of nitric oxide in biological tissues. Nat Med 1:804–809

66. Kuppusamy P, Wang P, Samouilov A, Zweier JL (1996) Spatial mapping of nitric oxide generation in the ischemic heart using electron paramagnetic resonance imaging. Magn Reson Med 36:212–218

67. Yoshimura T, Yokoyama H, Fujii S, Takayama F, Oikawa K, Kamada H (1996) In vivo EPR detection and imaging of endogenous NO in LPS-treated mice. Nat Biotechnol 14:992–994

68. Fujii H, Koscielniak J, Berliner LJ (1997) Determination and characterization of nitric oxide generation in mice in vivo L-band EPR spectroscopy. Magn Res Med 38:565–568

69. Mikoyan VD, Kubrina LN, Serezhenkov VA, Stukan RA, Vanin AF (1997) Complexes of Fe2+ with diethyldithiocarbamate or N-methyl-D-glucamine dithiocarbamate as traps of nitric oxide in animal tissues: comparative investigations. Biochim Biophys Acta 1336:225–234

70. Kotake Y, Moore DR, Sang H, Reinke LA (1999) Continuous monitoring of in vivo nitric oxide formation using EPR analysis in biliary flow. Nitric Oxide 3:114–122

71. Tsuchiya K, Jiang JJ, Yoshizumi M, Tamaki T, Houchi H, Minakuchi K, Fukuzawa K, Mason RP (1999) Nitric oxide-forming reactions of the water-soluble nitric oxide spin-trapping agent, MGD. Free Radic Biol Med 27:347–355

72. Lu C, Koppenol WH (2005) Redox cycling of iron complexes of N-(dithiocarboxy)sarcosine and N-methyl-d-glucamine dithiocarbamate. Free Radic Biol Med 39:1581–1590

73. Tsuchiya K, Takasugi M, Minakuchi K, Fukuzawa K (1996) Sensitive quantitation of nitric oxide by EPR spectroscopy. Free Radic Biol Med 21:733–737

74. McGrath C, O'Connor C, Sangregodo C, Seddon J, Sinn E, Sowrey E, Young N (1999) Direct measurement of the high-spin and low-spin bond lengths and the spin state population in mixed spin state systems: an Fe K-edge XAFS study of iron(III) dithiocarbamate complexes. Inorg Chem Commun 2:536–539

75. Tsuchiya K, Yoshizumi M, Houchi H, Mason R (2000) Nitric oxide-forming reaction between the iron-N-methyl-D-glucamine dithiocarbamate complex and nitrite. J Biol Chem 275:1551–1556

76. Kubrina LN, Mikoyan VD, Mordvintcev PI, Vanin AF (1993) Iron potentiates lipopolysaccharide-induced nitric oxide formation in animal organs. Biochim Biophys Acta 1176:240–244

77. Vanin AF, Huisman A, Stroes ESG, Ruijter-Heijstek FC, Rabelink TJ, Faassen EE (2001) Antioxidant capacity of mononitrosyl-iron-dithiocarbamate complexes: implications for NO trapping. Free Radic Biol Med 30:813–824

78. Tsuchihashi K, Kirima K, Yoshizumi M, Houchi H, Tamaki T, Mason RP (2002) The role of thiol and nitrosothiol compounds in the nitric oxide-forming reactions of the iron-N-methyl-D-glucamine dithiocarbamate complex. Biochem J 367:771–779

79. Stamler JS, Singel DJ, Loscalzo J (1992) Biochemistry of nitric oxide and its redox-activated forms. Science 258:1898–1902

80. Komarov AM, Wink DA, Feelisch M, Schmidt HHW (2000) Electron-paramagnetic resonance spectroscopy using N-methyl-D-glucamine dithiocarbamate iron cannot discriminate between nitric oxide and nitroxyl: implications for the detection of reaction products for nitric oxide synthase. Free Radic Biol Med 28:739–742

81. Xia Y, Cardounel AJ, Vanin AF, Zweier JL (2000) Electron paramagnetic resonance spectroscopy with N-methyl-D-glucamine dithiocarbamate iron complexes distinguishes nitric oxide and nitroxyl anion in a redox-dependent manner: applications in identifying nitrogen monoxide products from nitric oxide synthase. Free Radic Biol Med 29:793–797

82. Cocco D, Calabrese L, Rigo A, Agrese E, Rotilio G (1981) Re-examination of the reaction of diethyldithiocarbamate with the copper of superoxide dismutase. J Biol Chem 256:8983–8986

Chapter 9

Measurement of Nitric Oxide (NO) Generation Rate by Chloroplasts Employing Electron Spin Resonance (ESR)

Andrea Galatro and Susana Puntarulo

Abstract

Chloroplasts are among the more active organelles involved in free energy transduction in plants (photophosphorylation). Nitric oxide (NO) generation by soybean (*Glycine max*, var ADM 4800) chloroplasts was measured as an endogenous product assessed by electron paramagnetic resonance (ESR) spin-trapping technique. ESR spectroscopy is a methodology employed to detect species with unpaired electrons (paramagnetic). This technology has been successfully applied to different plant tissues and subcellular compartments to asses both, NO content and generation. The spin trap MGD-Fe^{2+} is extensively employed to efficiently detect NO. Here, we describe a simple methodology to asses NO generation rate by isolated chloroplasts in the presence of either L-Arginine or nitrite (NO_2^-) as substrates, since these compounds are required for enzymatic activities considered as the possible sources of NO generation in plants.

Key words ESR, Chloroplasts, NO generation, NOS*like* activity, NO_2^-

1 Introduction

Electron spin resonance (ESR) spectroscopy is a methodology employed to measure species with unpaired electrons (paramagnetic) [1], such as free radicals and transition metals. Paramagnetism is related to the magnetic moment exerted by the unpaired electron and allows the use of ESR in a wide array of experimental conditions. This high degree of selectivity renders ESR useful even when working with complex biological systems [1].

NO is an inorganic free radical gaseous molecule with multiple roles in biological systems [2]. NO is itself paramagnetic with the free electron being shared between N_2 and O_2. The broader chemistry of NO involves a redox array of species with distinctive properties and reactivities: NO^+ (nitrosonium), NO^- (nitroxyl anion), and NO (NO radical). Neutral NO has a single electron in its 2p-π antibonding orbital and the removal of this electron forms NO^+ while the addition of one more electron to NO forms NO^- [3]. From a biological point of view the important reactions of NO are

Kapuganti Jagadis Gupta (ed.), *Plant Nitric Oxide: Methods and Protocols*, Methods in Molecular Biology, vol. 1424,
DOI 10.1007/978-1-4939-3600-7_9, © Springer Science+Business Media New York 2016

those with O_2 and its various redox forms and with transition metal ions. NO also reacts rapidly with O_2^- in aqueous solution, yielding peroxynitrite ($ONOO^-$) [4].

When discussing the chemistry and physiological effects of NO, it should be considered that NO is a highly diffusible second messenger that can elicit effects relatively far from its site of production. The concentration and therefore the source of NO are the major factors determining its biological effects [5]. At low concentrations (<1 μM) the direct effects of NO predominate. At higher concentrations (>1 μM), the indirect effects mediated by reactive nitrogen species (RNS) prevail. The direct effects of NO involve the interaction of NO with metal complexes or proteins leading to tyrosine nitration, selectively and reversibly, and it has been shown that there are $ONOO^-$ dependent and independent pathways for the nitration in vivo [6]. NO also is able to terminate lipid peroxidation [7]. The indirect effects of NO, produced through the interaction of NO with either O_2 or O_2^-, include nitrosation (when NO^+ is added to an amine, thiol, or hydroxy aromatic group), oxidation (when one or two electrons are removed from the substrate), or nitration (when NO_2^+ is added to a molecule) [5]. $ONOO^-$ acts as both, nitrating agent and powerful oxidant, to modify proteins (formation of nitrotyrosine), lipids (lipid oxidation, lipid nitration), and nucleic acids (DNA oxidation and DNA nitration) [8]. In summary, the potential reactions of NO are numerous and depend on many different factors. Thus, the relative balance between oxidative and nitrosative stress should be carefully evaluated for better understanding the complexity of NO biological effects.

To detect NO, different methodologies have been developed employing spectrophotometric [9] and fluorescent techniques [10], an O_2 monitor, or ESR [11]. The O_2 monitor methodology was developed to measure the consumption of O_2 in liquid phase by NO. This method, often used for quantification of aqueous stock solutions of NO, is based on the reaction of NO with O_2, according to the reaction 1, where the consumption of O_2 is recorded.

$$4NO + O_2 + 2H_2O \rightarrow 4HNO_2 \qquad [1]$$

ESR is one of the most powerful techniques for the detection and identification of biological radicals, being certainly the only method by which NO and its paramagnetic derivatives can be unambiguously identified [12]. Previous work has shown the capacity of ESR of detecting NO in the presence of exogenous traps in soybean embryonic axis [13] and cotyledons [14, 15], or sorghum embryonic axes [16]. Although no nitric oxide synthase (NOS) enzyme has been identified in plants, a NOS*like* activity has been extensively reported. Caro and Puntarulo [13] have determined a NADPH-diaphorase activity in homogenates from soybean axes. Galatro et al. [17] and Jasid et al. [18] have assessed L-Argi-nine (L-Arg) dependent NO generation by soybean leaves, and soybean chloroplasts employing ESR. Moreover, NO generation by L-Arg-dependent NOS activity was described in isolated peroxi-somes from pea leaves employing ESR [19].

Chloroplasts are key organelles in plant metabolism and they seem to be involved in NO production [15, 20–23]. Two independent pathways for NO generation in isolated chloroplasts from soybean plants have been described: (i) one dependent on L-Arg and NADPH (NOS*like*), and (ii) another dependent on nitrite (NO_2^-) [18] (Fig. 1). These NO generation sources were evaluated employing the spin trap sodium-*N*-methyl-D-glucamine dithiocarbamate $(MGD)_2$-Fe^{2+} (Fig. 2), and the required substrates and cofactors [18]. Galatro et al. [15] showed that chloroplasts contribute to NO synthesis in vivo employing both, confocal fluorescence microscopy, and EPR techniques. The level of NO in the soybean cotyledons was related to chloroplasts functionality. The detection of NO in coincidence with cotyledon maximum fresh weight, chlorophyll content, and quantum yield of PSII, supported the hypothesis of a strong link between NO levels and chloroplast functionality. Moreover, seedlings exposed *in vivo* to herbicides showed an impaired NO accumulation, and deleterious effects on chloroplast function (loss of photosynthetic capacity). The use of the herbicide DCMU (3-(3,4-dichlorophenyl)-1,1-dimethyl urea), that binds plastoquinone and blocks electron flow at the quinone acceptors of photosystem II, supports a role for the integrity of the photosynthetic electron chain in chloroplasts NO production *in vivo*, as has been previously observed by Jasid et al. [18] in isolated soybean chloroplasts.

In this chapter we describe a simple ESR methodology to asses NO generation rate by isolated chloroplasts in the presence of either L-Arg or NO_2^-.

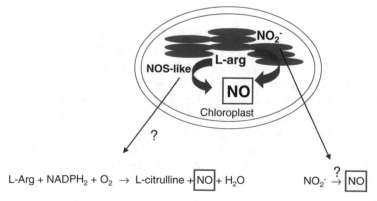

Fig. 1 Scheme of the main proposed sources of NO in chloroplasts. NOS*like*, nitric oxide-like activity, and NO_2^--dependent NO generation

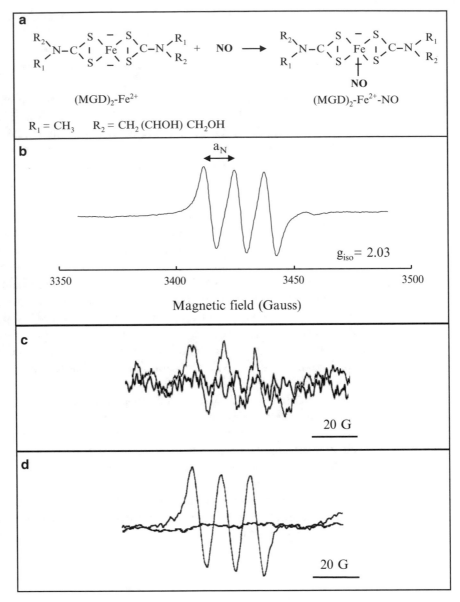

Fig. 2 ESR-spin trapping detection of NO. (**a**) Chemical reaction between (MGD)$_2$-Fe^{2+} complex and NO to form (MGD)$_2$-Fe^{2+}-NO adduct. (**b**) Typical EPR spectra of (MGD)$_2$-Fe^{2+}-NO adduct obtained employing a standard solution of 0.1 mM S-nitrosoglutathione (GSNO) mixed with the spin trap. (**c**) L-Arg-dependent NO generation in isolated chloroplasts: ESR spectra of the (MGD)$_2$-Fe^{2+}-NO adduct from soybean chloroplasts incubated for 10 min in the presence of MGD-Fe (10:1 mM), 1 mM Arg, 1 mM CaCl$_2$, 5 mM MgCl$_2$, and 0.1 mM NADPH, overlapped to the spectrum obtained with the reaction media in absence of chloroplasts. (**d**) NO$_2^-$-dependent NO generation by isolated chloroplasts: ESR spectra of the (MGD)$_2$-Fe^{2+}-NO adduct from soybean chloroplasts incubated for 3 min in the presence of MGD-Fe (10:1 mM) and 1 mM NaNO$_2$ overlapped to the spectrum obtained with the reaction media in absence of chloroplasts. Taken and modified from Puntarulo et al. [29] and Jasid et al. [18]

2   Materials

2.1   Chloroplasts Isolation Reagents

2.1.1   Isolation Buffer

50 mM HEPES (*N*-[2-Hydroxyethyl] piperazine-*N'*-[2-ethane-sulfonic acid]), pH 7.6, 330 mM sorbitol, 2 mM EDTA (Ethylenediaminetetraacetic acid), 1 mM $MgCl_2$, 0.05 % (w/v) bovine serum albumin (BSA), 5 mM ascorbic acid, and protease inhibitors (40 μg/mL phenylmethylsulfonyl fluoride (PMSF), 0.5 μg/mL leupeptin, 0.5 μg/mL aprotinin). To prepare 500 mL (final volume) weigh 5.958 g of HEPES, 30 g of sorbitol, 0.38 g of EDTA, 0.250 g of BSA, and 9.52 mg of $MgCl_2$. Add 400 mL of distilled water, mix employing a magnetic stirrer, and adjust the pH to 7.6 employing 1 M NaOH solution. Complete the volume to 500 mL.

This buffer can be stored at 4 °C. PMSF, protease inhibitors and ascorbic acid must be added previous to chloroplast isolation (*see* **Note 1**).

2.1.2   Wash and Resuspension Buffer

50 mM HEPES, pH 8.0, 330 mM sorbitol. To prepare 200 mL of buffer weight 2.383 g of HEPES, 12 g of sorbitol, add 150 mL of distilled water, mix employing a magnetic stirrer and adjust the pH to 8.0 employing 1 M NaOH solution. Complete the volume to 200 mL. Store the buffer at 4 °C.

2.1.3   Percoll Cushion

50 mM HEPES, pH 8.0, 330 mM sorbitol, 30 % [v/v] Percoll. To 12 mL of Percoll add 50 mM HEPES buffer, pH 8.0, 330 mM sorbitol (wash and resuspension buffer) up to 40 mL final volume.

2.2   NO Detection Reagents

2.2.1   Spin Trap Solution

100 mM potassium phosphate buffer pH 7.4, containing the spin trap (20 mM MGD, 2 mM $FeSO_4$). The suspension of isolated chloroplast will be supplemented with equal volume of 100 mM potassium phosphate buffer pH 7.4 containing the spin trap. Weight MGD (MW 293.34, 7.3 mg/mL, 25 mM) and dissolve it in 100 mM potassium phosphate buffer, pH 7.4. To prepare a 10 mM $FeSO_4$ solution in 0.01 N HCl, weight 13.9 mg of $FeSO_4 \cdot 7H_2O$ (278.02 g/mol) and dissolve it in 5 mL of 0.01 N HCl. Add 200 μL 10 mM $FeSO_4$ to 800 μL MGD solution (final concentration will be 20 mM MGD, 2 mM $FeSO_4$, *see* **Note 2**).

2.2.2   L-Arg-Dependent NO Generation Rate

100 mM potassium phosphate buffer containing the spin trap (10 mM MGD, 1 mM $FeSO_4$), 1 mM L-Arg, 1 mM $CaCl_2$, 5 mM $MgCl_2$, 0.1 mM NADPH, and chloroplasts. Prepare the following concentrated solutions of L-Arg (50 mM, in 100 mM potassium phosphate buffer pH 7.4), $CaCl_2$ (50 mM, in distilled water), $MgCl_2$ (100 mM, in distilled water), and NADPH (10 mM, in distilled water). These solutions will be added to the buffer containing the chloroplasts at the moment of the measurement as it is described in the Method section (*see* **Note 3**). All the solutions can be stored at −20 °C, except for the NADPH solution that should be conserved at −70 °C.

2.2.3 NO$_2^-$-dependent NO generation rate	100 mM potassium phosphate buffer, pH 7.4, containing the spin trap (10 mM MGD, 1 mM FeSO$_4$), 1 mM NaNO$_2$, and the suspension of chloroplasts. Prepare a concentrated solution of NaNO$_2$ (100 mM, in distilled water). This solution will be added to the buffer containing the chloroplasts at the moment of the measurement, as it is described in the Method section.
2.3 ESR Equipment	Bruker ER 070 spectrometer (Karlsruhe, Germany) operating at 17–19 °C (room temperature).

3 Methods

3.1 Chloroplast Isolation

All procedures should be carried out at 4 °C in an ice bath.

1. Add 100 mL of isolation buffer to 10 g of soybean leaves. Homogenize the plant material in a blender employing short periods of blending (1 or 2 s) to disaggregate the tissue.

2. Filter the homogenate through two layers of Mira cloth.

3. Centrifuge the filtrate 5 min at $1500 \times g$ at 4 °C (40 mL tubes).

4. Discard the supernatant and resuspend the pellets gently in isolation buffer (1 mL).

5. Load this suspension (1 mL) in 10 mL of Percoll cushion and centrifuge at $4000 \times g$ for 10 min (*see* **Note 4**).

6. Collect the pellet containing the intact chloroplasts employing a Pasteur pipette, in 2 mL Eppendorf tubes (no more than 1 mL per Eppendorf), add wash and resuspension buffer up to 2 mL, and centrifuge for 5 min at $1500 \times g$ (4 °C). Suspend the pellet in 2 mL of the same buffer.

7. Repeat the washing procedure and suspend all the final pellets in the same buffer (i.e., 0.5 mL). Final protein and chlorophyll concentration will depend on the performance of the isolation procedure and the intactness of chloroplasts fraction obtained, as only intact chloroplast pass through the Percoll cushion (it can be obtained around 1–2 mg protein mL^{-1}) (*see* **Note 5**).

3.2 ESR Measurement of NO Generation Rate

3.2.1 L-Arg-Dependent NO Generation Rate

1. To asses L-Arg-dependent NO generation rate by chloroplasts, the organelles have to be subjected to osmotic shock in HEPES buffer 50 mM, pH 8.0, by the lack of sorbitol in the medium. The chloroplasts suspension (150 µL) should be centrifuged for 5 min at $1500 \times g$ (4 °C). Discard the supernatant and resuspend the chloroplasts in 150 µL of 50 mM HEPES buffer, pH 8.0. Mix this volume of membrane disrupted chloroplasts (1 mg protein mL^{-1}) with an equal volume of 100 mM phosphate buffer pH 7.4, containing the spin trap (20 mM MGD, 2 mM FeSO$_4$).

2. Add the appropriate cofactors: to 270 µL of chloroplasts suspension mixed with the spin trap solution, add 6 µL of Arg 50 mM, 15 µL of 100 mM $MgCl_2$, 6 µL 50 mM $CaCl_2$, and finally 3 µL 10 mM NADPH, and incubate up to 10 min at room temperature. The final concentration will be: 1 mM Arg, 1 mM $CaCl_2$, 5 mM $MgCl_2$, 0.1 mM NADPH in 300 µL of the suspension containing the chloroplasts and 10:1 mM (MGD-Fe) (*see* **Note 6**).

3. Transfer the sample to bottom-sealed Pasteur pipettes to record the EPR spectra at room temperature (18 °C) employing the following instrument settings: microwave frequency 9.5 GHz, 200 G field scan, 83.9 s sweep time, 328 ms time constant, 5.983 G modulation amplitude, 50 kHz modulation frequency, 20 mW microwave power, and 3400 G center field. The scan number must be adjusted according to the NO generation rate of the sample.

4. This procedure for assessing NO generation rate will be repeated at different time points (i.e., 10, 20, and 30 min) after starting the reaction by the addition of NADPH.

5. Quantification of the spin adduct could be performed using an aqueous solution of TEMPOL (4-hydroxy-2,2,6,6-tetramethylpiperidine 1-oxyl). The concentration of the $(MGD)_2$-Fe^{2+}-NO adduct is obtained by double integration of the three lines and cross-checked with the TEMPOL spectra. The amount of TEMPOL spins in the EPR cavity range from 0.5 to 5 nmol (standard curve) because 50 µL of a 10–100 µM solution of TEMPOL are added to the cavity employing the bottom-sealed Pasteur pipettes (*see* **Note 7**).

3.2.2 NO_2^--Dependent NO Generation Rate

1. Mix 150 µL of intact isolated chloroplasts (1 mg prot mL^{-1}) with equal volume of 100 mM potassium phosphate buffer pH 7.4 containing the spin trap (20 mM MGD, 2 mM $FeSO_4$). To 297 µL of this suspension, add 3 µL of 100 mM $NaNO_2$, and incubate up to 10 min at room temperature under ambient light conditions. The final concentration will be 1 mM $NaNO_2$, and 10:1 mM (MGD-Fe) in a final volume of 300 µL.

2. Transfer the sample to a bottom-sealed Pasteur pipette to record the EPR spectra at room temperature (18 °C) employing the instrument settings previously described (*see* **Note 8**).

This procedure for assessing NO generation rate will be repeated at different time points (i.e., 10, 20, and 30 min) after starting the reaction by the addition of $NaNO_2$.

4   Notes

1. Protease inhibitors and ascorbic acid must be added at the moment of performing the isolation procedure. Prepare a solution of 10 mg/mL of aprotinin, and 10 mg/mL of leupeptin in 100 mM potassium phosphate buffer, pH 7.4. Aliquots of protease inhibitors solutions (10 mg/mL) can be stored at −20 °C. To 100 mL of 50 mM HEPES, pH 7.6, 330 mM sorbitol, 2 mM EDTA, 1 mM $MgCl_2$, and 0.05 % (w/v) bovine serum albumin, add 5 μL of each protease inhibitor solution, 4 mg of PMSF, and 0.088 g of ascorbic acid. Mix the solution.

2. When MGD and Fe solutions are mixed the final solution turns brown. MDG and Fe solutions should be mixed immediately before the measurement. The MDG solution should be freshly prepared. However, the solution of 10 mM $FeSO_4$ in 0.01 N HCl could be previously prepared and stored at −20 °C.

3. NADPH concentration in the stock solution should be checked by measuring the absorbance at $\lambda = 340$ nm ($\varepsilon = 6.22$/mM/cm). To prepare a 10 mM NADPH solution dissolve 50 mg of NADPH (MW 833.35) in 6 mL of distilled water. The absorbance at $\lambda = 340$ nm should be recorded to assess the concentration. Aliquots of this solution could be stored at −70 °C.

4. Depending on both, the percoll gradient and the plant material, slight adjustments should be performed on the centrifugation speed (i.e., 4000–5000 × g), and time (10–12 min).

5. The protein content in the final suspension of chloroplasts could be measured according to Bradford [24]. The intactness of chloroplasts could be determined as ferricyanide-dependent O_2 evolution according to Edwards et al. [25]. The purity of the obtained fraction could be analyzed by biochemical assays, such as the hydroxypyruvate reductase activity for assessing peroxisomal contamination [26], the phosphoenolpyruvate carboxylase activity, a cytosolic marker [27], and the fumarase activity as a marker of mitochondrial presence [28].

6. To explore the efficiency of classic known NOS inhibitors on the obtained enzymatic activity the incubations should be performed in the presence of 5 mM L-NAME (N_ω-nitro-L-Arg methyl ester hydrochloride) or L-NNA (N_ω-nitro-L-Arg) since they are Arg analogs. Controls with boiled chloroplasts (exposed 20 min to 100 °C), in absence of NADPH, and in absence of chloroplasts should be added to the experimental protocol. The signal of the basal system, consisting in buffer containing the spin trap without any addition, should be measured under the same conditions.

7. TEMPOL is a stable free radical and may be used as a standard to quantify the free radical adduct generation rate. TEMPOL solutions are standardized spectrophotometrically at $\lambda = 429$ nm ($\varepsilon = 13.4/M/cm$ according to Jasid et al. [18]).

8. The inclusion in the experimental protocols of control samples without NO_2^-, or employing 1 mM $NaNO_3$ instead of 1 mM $NaNO_2$, boiled chloroplasts (20 min to 100 °C), and without chloroplasts must be considered. The supplementation with herbicides acting in chloroplasts, such as 3-(3,4 dichlorophenyl)-1,1- dimethyl urea (DCMU) (1 μM), may be employed to evidence if the intactness of electron transfer chain of chloroplasts is necessary for the NO generation (*see* ref [18]).

Acknowledgments

This study was supported by grants from the UBA, ANPCyT and CONICET. S.P. and A.G. are career investigators from CONICET.

References

1. Rosen GM (1985) Use of spin trap in biological systems. Free Radic Biol Med 1:345–375

2. Moncada S, Palmer RMJ, Higgs EA (1991) Nitric oxide: physiology, pathophysiology and pharmacology. Pharmacol Rev 43:109–142

3. Stamler JS, Singel DJ, Loscalzo J (1992) Biochemistry of nitric oxide and its redox-activated forms. Science 258:1898–1902

4. Saran M, Michel C, Bors W (1990) Reaction of NO with O_2. Implications for the action of endothelium-derived relaxing factor (EDRF). Free Radic Res Commun 10:221–222

5. Wink DA, Mitchell JB (1998) Chemical biology of nitric oxide: insights into regulatory, cytotoxic, and cytoprotective mechanisms of nitric oxide. Free Radic Biol Med 25:434–456

6. Davis KL, Martin E, Turko IV et al (2001) Novel effects of nitric oxide. Annu Rev Pharmacol Toxicol 41:203–236

7. Rubbo H, Parthasarathy S, Barnes S et al (1995) Nitric oxide inhibition of lipoxygenase-dependent liposome and low density lipoprotein oxidation: termination of radical chain propagation reactions and formation of nitrogen containing oxidized lipid derivatives. Arch Biochem Biophys 324:15–25

8. Gisone P, Dubner D, Pérez MR et al (2004) The role of nitric oxide in the radiation-induced effects in the developing brain. In Vivo 18:281–292

9. Grisham MB, Johnson GG, Lancaster JR Jr (1996) Quantitation of nitrate and nitrite in extracellular fluids. Methods Enzymol 268:237–246

10. Miles AM, Wink DA, Cook JC et al (1996) Determination of nitric oxide using fluorescence spectroscopy. Methods Enzymol 268:105–120

11. Ventakaraman S, Martin SM, Schafer FQ et al (2000) Detailed methods for the quantification of nitric oxide in aqueous solutions using either an oxygen monitor or EPR. Free Radic Biol Med 29:580–585

12. Jackson SK, Hancock JT, James PE (2007) Biological free radicals and biomedical applications of EPR spectroscopy. Electron Paramag Res 20:157–191

13. Caro A, Puntarulo S (1999) Nitric oxide generation by soybean embryonic axes. Possible effect on mitochondrial function. Free Radic Res 31:S205–S212

14. Jasid S, Galatro A, Villordo JJ et al (2009) Role of nitric oxide in soybean cotyledon senescence. Plant Sci 176:662–668

15. Galatro A, Puntarulo S, Guiamet JJ et al (2013) Chloroplast functionality has a positive effect on nitric oxide level in soybean cotyledons. Plant Physiol Biochem 66:26–33

16. Jasid S, Simontacchi M, Puntarulo S (2008) Exposure to nitric oxide protects against oxidative damage but increases the labile iron pool in sorghum embryonic axes. J Exp Bot 59:3953–3962

17. Galatro A, Simontacchi M, Puntarulo S (2004) Effect of nitric oxide exposure on antioxidant capacity of soybean leaves. Current Topics Plant Biol 5:69–79

18. Jasid S, Simontacchi M, Bartoli CG et al (2006) Chloroplasts as a nitric oxide cellular source. Effect of reactive nitrogen species on chloroplastic lipids and proteins. Plant Physiol 142:1246–1255

19. del Rio LA (2011) Peroxisomes as a cellular source of reactive nitrogen species signal molecules. Arch Biochem Biophys 506:1–11

20. Foissner I, Wendehenne D, Langebartels C et al (2000) In vivo imaging of an elicitor-induced nitric oxide burst in tobacco. Plant J 23:817–824

21. Gould KS, Lamotte O, Klinguer A et al (2003) Nitric oxide production in tobacco leaf cells: a generalized stress response? Plant Cell Environ 26:1851–1862

22. Arnaud N, Murgia I, Boucherez J et al (2006) An iron induced nitric oxide burst precedes ubiquitin-dependent protein degradation for Arabidopsis AtFer1 ferritin gene expression. J Biol Chem 281:23579–23588

23. Tewari RK, Prommer J, Watanabe M (2013) Endogenous nitric oxide generation in protoplast chloroplasts. Plant Cell Rep 32:31–44

24. Bradford MM (1976) A rapid and sensitive method for the quantitation of microgram quantities of protein utilizing the principle of protein-dye binding. Anal Biochem 72:248–254

25. Edwards GE, Lilley RMC, Craig S et al (1979) Isolation of intact and functional chloroplasts from mesophyll and bundle sheath protoplasts of C4 plant Panicum miliaceum. Plant Physiol 63:821–827

26. Schwitzguébel JP, Siegenthaler PA (1984) Purification of peroxisomes and mitochondria from spinach leaf by Percoll-density gradient centrifugation. Plant Physiol 75:670–674

27. Quy LV, Foyer C, Champigny ML (1991) Effect of light and NO_3^- on wheat leaf phosphoenolpyruvate carboxylase activity. Plant Physiol 97:1476–1482

28. Bartoli CG, Gómez F, Martínez DE et al (2004) Mitochondria are the main target for oxidative damage in leaves of wheat (*Triticum aestivum* L.). J Exp Bot 55:1663–1669

29. Puntarulo S, Jasid S, Boveris AD et al (2009) Electron Paramagnetic Resonance as a tool to study nitric oxide generation in plants. In: Hayat S, Mori M, Pichtel J, Ahmad A (eds) Nitric oxide in plant physiology. WILEY-VCH Verlag GmbH & Co. KGaA, Weinheim, pp 17–30

Chapter 10

Laser-Based Methods for Detection of Nitric Oxide in Plants

Julien Mandon, Luis A.J. Mur, Frans J.M. Harren, and Simona M. Cristescu

Abstract

Nitric oxide (NO) plays an important role in plant signaling and in response to various stress conditions. Therefore, real-time measurements of NO production provide better insights into understanding plant processes and can help developing strategies to improve food production and postharvest quality. Using laser-based spectroscopic methods, sensitive, online, *in planta* measurements of plant-pathogen interactions are possible. This chapter introduces the basic principle of the optical detectors using different laser sources for accurate monitoring of fast dynamic changes of NO production. Several applications are also presented to demonstrate the suitability of these detectors for detection of NO in plants.

Key words Laser, Optical detection, Nitric oxide, Plant-pathogen interaction

1 Introduction

Nitric oxide (NO) acts as a signaling molecule within species from every biological kingdom. Its utility as a signal arises from the physical properties of NO. Nitric oxide is a remarkably mobile molecule being able to move freely across membranes and, with very low solubility in aqueous solutions (partition coefficient: $0.04 = 4.6$ ml/100 ml H_2O @ 20 °C, 1 atm.), it can readily enter the gaseous phase. As NO can react with atmospheric O_2 producing NO_2, the theoretical half-life is about 6 s; however under physiological conditions, this is likely to be extended to ~30 s [1]. The NO molecule exists in a variety of reduced states due to the possession of an unpaired electron, such as –NO (nitroxyl ion), NO, and +NO (nitrosonium ion), which are reactive nitrogen intermediates (RNIs). These RNIs allow specific interactions with, for example, proteins, to fulfil its regulatory functions [2].

After extensive characterization in animal systems, NO has been shown to function as a signal in plant development and in responses to stress. These functions have been extensively reviewed

Kapuganti Jagadis Gupta (ed.), *Plant Nitric Oxide: Methods and Protocols*, Methods in Molecular Biology, vol. 1424, DOI 10.1007/978-1-4939-3600-7_10, © Springer Science+Business Media New York 2016

in several recent papers [3–6] and range from the single cellular responses of stomatal guard cells to large organ development such as roots and flowers. In postharvest technology, early exposure (2–6 weeks) of lettuce plants to NO slowed growth compared to controls [7]. However, NO treatments after 8 weeks up to the harvest time (10 weeks) increased the lettuce growth rates as indicated by leaf area and net assimilation rates. The germination of lettuce was also stimulated by imbibing seeds in water containing a NO-generating chemical ("NO donor") [8]. Beyond lettuce, fumigation of several fruit species (strawberry, cucumber, kiwi) with 50–250 ppb of NO for 2–16 h results in a longer postharvest lifetime up to 150 % [9].

Taken together, these observations suggest that accurate measurements of NO *in planta* and ideally online NO production are essential for a proper understanding of plant processes and to improve food production and postharvest quality.

2   Basic Principles of Optical Sensors

Several methods for *in* and *ex planta* NO detection have been developed and used. They have been extensively reviewed in Mur et al. [10]. Among them, optical detectors using laser sources in the mid-infrared (MIR) region allow online and sensitive detection of NO *in planta*.

For decades, optical laser-based sensors have been known to be efficient tools in the detection of molecules such as NO in gas samples. However, the performance of light sources in the specific MIR region considerably slowed down their industrial exploitation. Only recently, progress in semiconductor lasers and detectors triggered optical sensors in providing the reliability which would allow their use outside the physics laboratory environment. Continual research and development has led to the emergence of sensors with improved features in terms of response time, selectivity to a specific compound, sensitivity, and compactness.

The basic principle of a laser-based detector for NO monitoring is the detection of changes in light intensities or polarization when the laser light is interacting with NO molecules. As illustrated in Fig. 1, such detector includes three main elements: (1) the laser source which produces the light interacting with the NO molecules; (2) the gas cell containing the gas sample where light intensity is attenuated (due to absorption) and/or its polarization changes; and (3) the detection system able to sense these modifications (e.g., photodetector) to measure variation of light intensity.

The choice for the laser source is determined by its interaction with the NO molecules. The electromagnetic spectrum in the infrared wavelength region is often referred to as the "fingerprint" region. This designation reflects the strong and distinct

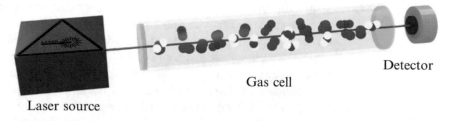

Fig. 1 A typical laser-based sensor. It is composed of a laser light source, a gas cell containing the sample to analyze, and a light detector (e.g., photodiode). The light is absorbed or its polarization is changed while passing through the gas sample. These changes are proportional to the concentration of the gas inside the cell and are measured with a detector

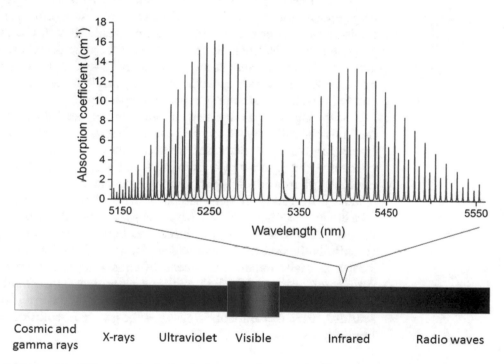

Fig. 2 Fingerprint of NO molecules in the electromagnetic spectrum. The strongest absorption features of NO are located in the mid-infrared around 5.3 μm

rotational-vibrational absorption bands of the vast majority of molecular species in this region from 2 to 20 μm. As such, this wavelength region is unique in its potential to identify and detect, with a high specificity and sensitivity, molecular species. In order to reach the highest sensitivity, the light source of the sensor should emit light where the interaction with the molecule is the most intense. For NO, the strongest absorption band is centered in the MIR region at 5.3 μm (covering a range from 5.1 to 5.7 μm), as shown in Fig. 2.

Until the 1990s, only three types of bulky and user-unfriendly lasers were suitable for NO detectors: CO lasers [11, 12], sources based on nonlinear optical parametric frequency conversion [13–15], and IV-VI lead-salt diodes [16, 17]. Therefore, other nonoptical based sensors came to dominate the market. Among them, chemiluminescence-based NO analyzers became the "gold" standard system in NO assessment and were extensively used in environmental research [18–23] and the measurement of exhaled NO in human breath [24–27]. The laser-based systems were at that stage unable to deliver sufficient high sensitivity in combination with fast response time that could be developed into a commercially viable sensor.

Further developments in the fabrication of MIR sources opened new opportunities for compact and stand-alone NO sensors. A new semiconductor laser source, the quantum cascade laser (QCL), was proposed in 1994 [28]. Soon these lasers became commercially available sources for almost any mid-infrared wavelength, offering good performances for spectroscopy, i.e., single-mode operation and fast wavelength scanning at room temperature.

Following the development of light sources, several detection methods for NO have been proposed. Taking advantage of the laser's features, the detection method consists of sensing the changes of the light when interacting with the studied sample. For example, photoacoustic spectroscopy consists of detecting the sound wave created when the sample is "heated" by the laser beam. The design of a photoacoustic based sensor can be very robust and relatively simple to set up, but it requires high-power light source to sufficiently excite the molecules. Another detection method known as cavity enhanced consists of enhancing the interaction path length or time of the light with the gas sample. The advantage is that no high-power laser is required; however they appear to be sensitive to environmental changes (e.g., temperature, and humidity).

3 Laser-Based Photoacoustic Detectors

The laser-based photoacoustic detectors are using the photoacoustic effect (PA) that consists of generation of characteristic acoustic waves by certain molecules as they absorb light (Fig. 3). The absorbed energy from light is transformed into kinetic energy of the sample by energy exchange processes. This results in local heating of the sample. Since the excited molecules dissipate thermal energy very quickly via relaxation processes (e.g., by collision with other molecules), the modulation of the light source at a certain frequency in the acoustic range results in rapid heat/cooling changes. This gives rise to a periodical pressure change, i.e., sound

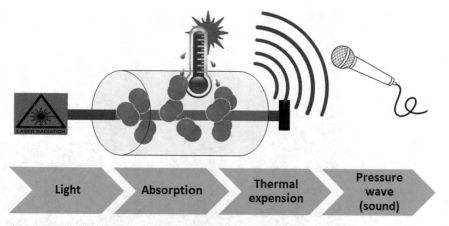

Fig. 3 Schematic view of a photoacoustic laser-based sensor. The molecules absorb the energy from the laser light and a pressure wave (sound wave) is generated due to their thermal expansion. The amplitude of this sound wave, proportional to the amount of molecules, is detected by a sensitive microphone

waves at the same frequency as this modulation. The amplitude of the sound wave created in this way is proportional to the concentration of the probed molecules and can be detected with sensitive microphones. Thus, absorbed light energy is detected as an acoustical signal (sound wave), hence the name photoacoustic effect.

In the photoacoustic effect, the intensity of the detected signal due to NO absorption is directly proportional with the laser power. Therefore a high laser power is a prerequisite for achieving high sensitivity of NO detection. The carbon monoxide (CO) laser was the first laser source to be combined with the photoacoustics for NO detection [29] in various applications ranging from environmental pollution to plant-pathogen interaction [30].

The CO lasers were demonstrated for the first time in 1965 [11]. The coherent light is produced by generating an electric discharge through the gas, i.e., CO. These lasers have been of great interest in physics research in the past. They could generate light in the MIR fingerprint region, at very high output power (several watts), providing single-mode operation (important for spectroscopy to be selective to the detection of specific molecule) and a very high beam quality. Considering spectroscopic applications, the CO laser emission spectra coincide with a large number of absorption features of relevant organic and inorganic compounds. Therefore, CO lasers have been employed to monitor trace gases in combination with a photoacoustic absorption cell [29, 31–35]. However, although CO lasers are very efficient sources for gas sensing using the photoacoustic effect, their bulky size and high running costs preclude their use; other laser source became more attractive for spectroscopic purposes in the same wavelength region. Figure 4 illustrates the size of a CO laser together with two

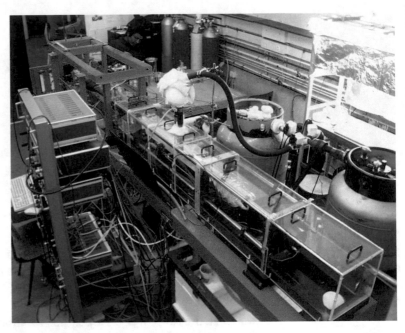

Fig. 4 Lab view of a CO-laser photoacoustic detector.(courtesy of Trace Gas Facility)

large cooling reservoirs with liquid nitrogen. Moreover, the use of CO lasers was limited by the extreme toxicity of CO gas. Care is needed for human operators in building and working with this deadly gas that is also extremely corrosive to many materials.

4   Quantum Cascade Laser-Based Detectors

QCLs were developed some decades after CO lasers [28]. QCLs are semiconductor lasers providing light in the MIR emission (between 4 and 20 μm), including the NO absorption band. The wavelength (or "color") of the light emitted by QCLs has to be specified before the design of the laser chip. It can be very selective to target a known molecule of interest; in this case the QCL is called distributed feedback-QCL (DFB-QCL). Another option is that the wavelength of the emitted light is widely tunable, such as in an EC-QCL (external cavity-QCL). The potential of QCL in the design of a compact gas sensor was rapidly recognized as they are extremely small devices (Fig. 5) and often can operate at room temperature. Nowadays, QCLs are fully capable to be integrated into an autonomous gas sensor working in field conditions, including for the sensitive detection of NO [36].

The main drawback of QCLs is the limited optical power, i.e., only few milliwatts in average, which is several orders of magnitude lower than of the CO laser. As a consequence, photoacoustic

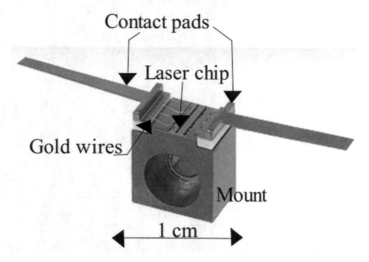

Fig. 5 A quantum cascade laser on its mount. The small size of QCLs makes them suitable for compact gas sensor design. Gold wires are used to supply electric current to the chip

technique is not appropriate, since the sensitivity of the sensor, i.e., the ability to detect very low amount of molecules, scales with the laser power. Different configurations have been associated with QCLs to achieve high sensitivity for NO detection, such as the use of a multipass gas cell [37–40], cavity-enhanced methods [37, 41–44], or the Faraday effect [45–48].

For laser-based sensors, it is well known that measurement sensitivity can be improved as the path length of a gas absorption cell increases. To do so, two mirrors are placed at both ends of a gas cell. Therefore, once injected into the gas cell, the light bounces several times between the two mirrors before exiting. As a result, in a gas cell of 30 cm long, the light can travel few hundred meters, as shown in Fig. 6. The differences between a multipass cell and cavity-enhanced-methods rely on the technology of the mirrors, metallic coating or dielectric, respectively, and the way the light is injected into the cell. Dielectric mirrors provide higher reflectivity, as such the light can travel up to several kilometres within the gas cell providing extreme sensitivity. The system is a bit more complex to handle than just using metallic mirrors.

Figure 7 illustrates a typical NO detector using QCL. It consists of a QCL placed in a housing for temperature stability, the gas cell with two highly reflective mirrors, and the infrared detector measuring the transmitted light through the cell. The amount of transmitted light is giving an information about the absorbed light by the sample, which is proportional to the concentration of the NO molecules present in the sample.

Fig. 6 Photo of a multipass gas cell. The laser light (*in red*) is injected into the cell and will undergo multiple reflections in between two mirrors in order to enhance its interaction with the NO gas inside the cell. Within a compact 30 cm long cell, the light can travel several hundreds of meters

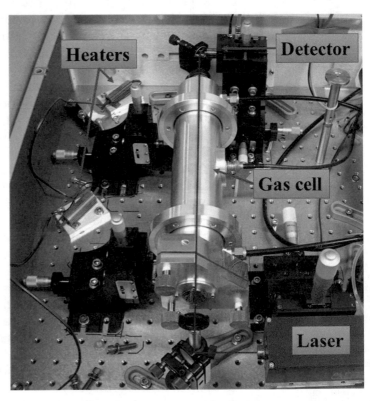

Fig. 7 A typical multipass cell setup combined with a QCL laser for the detection of NO (courtesy of Trace Gas Facility). The setup contains the laser emitting at 5.3 μm, the gas cell, and a light intensity detector. The setup is placed in a box controlled in temperature by thermistors (heaters). In *red* is the path of the infrared laser beam

5   Dynamic Nitric Oxide Measurements

Measuring NO produced by a biological system represents a challenge. Most of the time, the rate of gas production is very low and can vary over time. The gas sensor should be sensitive and quick enough to reveal the dynamics of the measured process. Optical sensors allow real-time continuous monitoring of NO emission over hours or days. For this, a flow-through system that consists of controlled gas flow over the cuvettes containing the plant material and automatic switching between the cuvettes can be used (Fig. 8).

Collection of the gases emitted by the plants without introducing additional stresses is not trivial. Different designs for the sealed vials can be considered for the headspace (Fig. 9). The entire plant can be placed inside a glass container or part of the plant attached or detached from the plant (e.g., leaves, fruits) can be isolated into specially designed cuvettes. Air is usually used as carrier gas at low flow rates over the plant material (max. 5 L/h) as needed in plant biology. The possibility to use these flow rates represents an advantage for the optical detectors as compared to chemiluminescence detectors that require sampling at higher flow rate and thus being unable to detect low NO emission rates.

The author's laboratories have been in collaboration for many years and have published extensively to provide many definitive

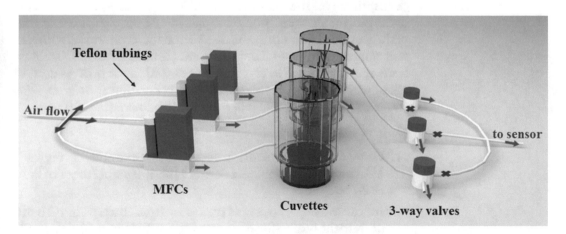

Fig. 8 Dynamic sampling. The cuvettes containing the biological material are continuously flushed with air at a constant flow rate (typically between 1 and 3 l/h) regulated by mass flow controllers (MFCs). The NO molecules emitted in the headspaces are transported successively by using electronic valves (three-way valves) to the gas cell of the laser-based sensor

Fig. 9 Examples of glass cuvette design for measuring NO emitted from different plant materials

descriptions of the patterns of NO production during plant-pathogen interactions. For example, in response to the bacterial pathogen, *Pseudomonas syringae* [30] or the economically important fungal pathogen—*Botrytis cinerea* (Mur et al. 2012). To illustrate, the value of the laser-based NO detection the process used to measure NO from *Pseudomonas syringae* pathovar *tomato* (*Pst*) strain DC3000 *avrRpm1* will be considered here (Fig. 10).

Protocol

1. Grow Arabidopsis plants in compost (for example Levington's M1 compost) under short day (8 h) at 200 μmol/m²/s of light at 20 °C. Water daily.

2. Inoculate a single colony of the *Pst avrRpm1* bacteria into 10 mL of nutrient broth (for example, Oxoid, UK) in a 50 mL sterile tube using standard aseptic techniques. Incubate the inoculated broth in an orbital shaker set to 50 rpm and at 28 °C. After 24 h the culture should have grown so to appear cloudy and the culture should be just entering its stationary phase.

3. Centrifuge the culture in a 50 mL sterile tube at $1000 \times g$ and at 10 °C to prevent the bacteria becoming too hot during

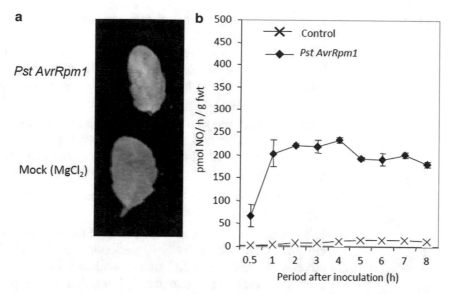

Fig. 10 Measuring NO production from Arabidopsis leaves challenged with *Pseudomonas syringae* pathovar tomato avrRpm1. (**a**) Phenotypes at 48 h post-inoculation with *Pseudomonas syringae* pv. tomato DC3000 (Pst) AvrRpm1 eliciting the form of programmed cell death known as the hypersensitive response (HR) or 10 mM MgCl₂ (considered to be a mock-inoculated control). (**b**) NO production was determined from Arabidopsis Col-0 an, following inoculation Pst AvrRpm1 or 10 mM MgCl₂ using a quantum cascade laser-based system (adapted from data published in Mur et al. [2012] JXB 63(12): 4375–87)

the centrifugation step and losing viability. Discard the supernatant.

4. Resuspend the bacterial pellet in 10 mL of 10 mM MgCl₂ to wash the pellet in an isotonic solution. Centrifuge the 50 mL tube as in **step 3** discarding the supernatant.

5. Resuspend the pellet in 10 mL of 10 mL MgCl₂.

6. Take 100 μL of the bacterial suspension and add 400 μL 10 mM MgCl₂ and measure the absorbance (turbidity) of the suspension in a spectrophotometer set to measure absorbance at 600 nm. Use 10 mM MgCl₂ as a blank to determine the increased absorbance in the bacterial suspension. At 600 nm, a bacterial population of 2×10^7 cells per mL has an absorbance of 0.1. Based on the absorbance reading dilute your original bacterial suspension in 10 mM MgCl₂ to give a bacterial population of 1×10^6 cells per mL.

7. To inoculate the Arabidopsis plants with bacteria use the barrel of a needle-less 1 mL syringe with the bacterial suspension by pulling back the plunger. Once filled push the plunger forward until there is no air in the barrel of the needle. This will avoid flooding the plant leaf with air.

8. Take the leaf and turn it so that the lowermost side is exposed. Do not detach the leaf. Carefully place the mouth of the syringe flush against the lower laminar (abaxial) of the leaf. It helps if the finger is placed underneath the syringe contact point, i.e., against the upper leaf (adaxial) laminar. The finger here acts as a cushion.

9. Gently squeeze the plunger under the infracellular spaces with the leaf filled with liquid. Leave the plant for 1 h to allow the inoculated liquid to disperse as a result of the plant transpiration stream.

10. Repeat **steps 8** and **9** on a separate plant and with 10 mM $MgCl_2$ to give an inoculation control.

11. It is recommended that all of the leaves of the Arabidopsis rosette are inoculated with the bacterial suspension or 10 mM $MgCl_2$ to increase the amount of NO that is produced. Prior trial inoculations of leaves are recommended so that the experimenter becomes skilled in the rapid infiltration of Arabidopsis leaves.

12. When the whole of the Arabidopsis rosette is inoculated carefully detach this from the roots at soil level. The rosettes are then enclosed within a cuvette as shown in Fig. 8 so that measurement could commence. The mass of plant material used was in the range of 4–6 g per cuvette.

13. A typical experimental setup will involve at least two cuvettes (one inoculated which are continuously flushed with air at a constant flow rate (typically between 1 and 3 l/h) regulated by mass flow controllers (MFCs)). Timed switching using electronic valves (three-way valves) allows the NO molecules emitted into the headspaces of a given cuvette to be carried into the gas cell of the laser-based sensor.

14. As stated above, the QCL emitting around 1900 cm^{-1} passes through an absorption multipass cell in which enters the airline transporting the NO released by a single inoculated rosette within a glass cuvette (~500 mL volume). The NO production is directly detected by measuring the attenuation of the laser intensity due to the NO absorption in the cell.

15. Typical measuring periods of NO production from a given cuvette are 20–30 min before switching to another cuvette.

16. This basic system can be modified to increase the number of cuvettes being measured as required by the experimenter. It is recommended that no more than four cuvettes be used as any increase in the number will increase the period required to cycle between each cuvette and reduce the number of NO measurements that may be taken.

17. Each experiment should be repeated at least once to ensure that consistent results are obtained.

6   Conclusion

In conclusion, laser-based detectors are suitable for real-time NO monitoring in various plant processes. The online, *in planta* measurements of plant-pathogen interactions provide a particularly important model for the utility of these systems as these experiments lead to dynamic changes in the rate of NO production. Clearly, the use of laser-based NO detected is not limited to plant-pathogen interactions and many other canonical NO measurements can be expected over the coming years.

References

1. Wink DA, Grisham MB, Mitchell JB et al (1996) Direct and indirect effects of nitric oxide in chemical reactions relevant to biology. Methods Enzymol 268:12–31
2. Gow AJ, Ischiropoulos H (2001) Nitric oxide chemistry and cellular signaling. J Cell Physiol 187(3):277–282
3. Domingos P, Prado AM, Wong A et al (2015) Nitric oxide: a multitasked signaling gas in plants. Mol Plant 8(4):506–520. doi:10.1016/j.molp.2014.12.010
4. Mur LAJ, Mandon J, Persijn S et al (2013) Nitric oxide in plants: an assessment of the current state of knowledge. AoB Plants. doi:10.1093/aobpla/pls052
5. Scheler C, Durner J, Astier J (2013) Nitric oxide and reactive oxygen species in plant biotic interactions. Curr Opin Plant Biol 16(4):534–539
6. Trapet P, Kulik A, Lamotte O et al (2015) NO signaling in plant immunity: a tale of messengers. Phytochemistry 112:72–79
7. Hufton CA, Besford RT, Wellburn AR (1996) Effects of NO (+NO2) pollution on growth, nitrate reductase activities and associated protein contents in glasshouse lettuce grown hydroponically in winter with CO2 enrichment. New Phytol 133(3):495–501
8. Beligni MV, Lamattina L (2000) Nitric oxide stimulates seed germination and de-etiolation, and inhibits hypocotyl elongation, three light-inducible responses in plants. Planta 210(2):215–221
9. Leshem YY, Wills RBH (1998) Harnessing senescence delaying gases nitric oxide and nitrous oxide: a novel approach to postharvest control of fresh horticultural produce. Biol Plant 41(1):1–10
10. Mur LAJ, Mandon J, Cristescu SM et al (2011) Methods of nitric oxide detection in plants: a commentary. Plant Sci 181(5):509–519
11. Patel CKN (1966) Vibrational-rotational laser action in carbon monoxide. Phys Rev 141(1):71–83
12. Osgood RM, Eppers WC (1968) High power Co-N2-He laser. Appl Phys Lett 13(12):409–411
13. Harris SE (1969) Tunable optical parametric oscillators. Proc IEEE 57(12):2096–2113
14. Giordmai JA, Miller RC (1965) Tunable coherent parametric oscillation in Linbo3 at optical frequencies. Phys Rev Lett 14(24):973–976
15. Baxter GW, Payne MA, Austin BDW et al (2000) Spectroscopic diagnostics of chemical processes: applications of tunable optical parametric oscillators. Appl Phys B 71(5):651–663
16. Partin DL (1985) Lead salt quantum well diode-lasers. Superlattice Microst 1(2):131–135
17. Tacke M (1995) New developments and application of tunable Ir lead salt lasers. Infrared Phys Technol 36(1):447–463
18. Clemitshaw KC (2004) A review of instrumentation and measurement techniques for ground-based and airborne field studies of gas-phase tropospheric chemistry. Crit Rev Env Sci Technol 34(1):1–108
19. Parrish DD, Fehsenfeld FC (2000) Methods for gas-phase measurements of ozone, ozone precursors and aerosol precursors. Atmos Environ 34(12-14):1921–1957
20. Kondo Y, Kawakami S, Koike M et al (1997) Performance of an aircraft instrument for the measurement of NOy. J Geophys Res Atmos 102(D23):28663–28671
21. Rockel P, Strube F, Rockel A et al (2002) Regulation of nitric oxide (NO) production by plant nitrate reductase in vivo and in vitro. J Exp Bot 53(366):103–110
22. Planchet E, Kaiser WM (2006) Nitric oxide (NO) detection by DAF fluorescence and chemiluminescence: a comparison using abiotic and biotic NO sources. J Exp Bot 57(12):3043–3055
23. Cristescu SM, Persijn ST, Hekkert STL et al (2008) Laser-based systems for trace gas detection in life sciences. Appl Phys B 92(3):343–349

24. Gustafsson LE, Leone AM, Persson MG et al (1991) Endogenous nitric-oxide is present in the exhaled air of rabbits, guinea-pigs and humans. Biochem Biophys Res Commun 181(2):852–857

25. Kharitonov SA, Yates D, Robbins RA et al (1994) Increased nitric-oxide in exhaled air of asthmatic-patients. Lancet 343(8890):133–135

26. Buchvald F, Baraidi E, Carraro S et al (2005) Measurements of exhaled nitric oxide in healthy subjects age 4 to 17 years. J Allergy Clin Immun 115(6):1130–1136

27. Ludviksdottir D, Janson C, Hogman M et al (1999) Exhaled nitric oxide and its relationship to airway responsiveness and atopy in asthma. Respir Med 93(8):552–556

28. Faist J, Capasso F, Sivco DL et al (1994) Quantum cascade laser. Science 264(5158): 553–556

29. Bernegger S, Sigrist MW (1990) Co-laser photoacoustic spectroscopy of gases and vapours for trace gas analysis. Infrared Phys 30(5): 375–429

30. Mur LAJ, Santosa IE, Laarhoven LJJ et al (2005) Laser photoacoustic detection allows in planta detection of nitric oxide in tobacco following challenge with avirulent and virulent Pseudomonas syringae pathovars. Plant Physiol 138(3):1247–1258

31. Bijnen FGC, Harren FJM, Hackstein JHP et al (1996) Intracavity CO laser photoacoustic trace gas detection: cyclic CH_4, H_2O and CO_2 emission by cockroaches and scarab beetles. Appl Opt 35(27):5357–5368

32. Zuckermann H, Harren FJM, Reuss J et al (1997) Dynamics of acetaldehyde production during anoxia and post-anoxia in red bell pepper studied by photoacoustic techniques. Plant Physiol 113(3):925–932

33. Martis AAE, Buscher S, Kuhnemann F et al (1998) Simultaneous ethane and ethylene detection using a co-overtone laser photoacoustic spectrometer: a new tool for stress/damage studies in plant physiology. Instrum Sci Technol 26(2-3):177–187

34. Dahnke H, Kahl J, Schuler G et al (2000) On-line monitoring of biogenic isoprene emissions using photoacoustic spectroscopy. Appl Phys B 70(2):275–280

35. Persijn ST, Veltman RH, Oomens J et al (2000) CO laser absorption coefficients for gases of biological relevance: H_2O, CO_2, ethanol, acetaldehyde, and ethylene. Appl Spectrosc 54(1):62–71

36. Wang Y, Nikodem M, Zhang E et al (2015) Shot-noise limited Faraday rotation spectroscopy for detection of nitric oxide isotopes in breath, urine, and blood. Sci Rep 5:9096

37. Menzel L, Kosterev AA, Curl RF et al (2001) Spectroscopic detection of biological NO with a quantum cascade laser. Appl Phys B 72(7): 859–863

38. Nelson DD, Shorter JH, McManus JB et al (2002) Sub-part-per-billion detection of nitric oxide in air using a thermoelectrically cooled mid-infrared quantum cascade laser spectrometer. Appl Phys B 75(2–3):343–350

39. Moeskops BWM, Cristescu SM, Harren FJM (2006) Sub-part-per-billion monitoring of nitric oxide by use of wavelength modulation spectroscopy in combination with a thermoelectrically cooled, continuous-wave quantum cascade laser. Opt Lett 31(6):823–825

40. McManus JB, Shorter JH, Nelson DD et al (2008) Pulsed quantum cascade laser instrument with compact design for rapid, high sensitivity measurements of trace gases in air. Appl Phys B 92(3):387–392

41. Kosterev AA, Malinovsky AL, Tittel FK et al (2001) Cavity ringdown spectroscopic detection of nitric oxide with a continuous-wave quantum-cascade laser. Appl Opt 40(30):5522–5529

42. Bakhirkin YA, Kosterev AA, Roller C et al (2004) Mid-infrared quantum cascade laser based off-axis integrated cavity output spectroscopy for biogenic nitric oxide detection. Appl Opt 43(11):2257–2266

43. Silva ML, Sonnenfroh DM, Rosen DI et al (2005) Integrated cavity output spectroscopy measurements of NO levels in breath with a pulsed room-temperature QCL. Appl Phys B 81(5):705–710

44. McCurdy MR, Bakhirkin Y, Wysocki G et al (2007) Performance of an exhaled nitric oxide and carbon dioxide sensor using quantum cascade laser-based integrated cavity output spectroscopy. J Biomed Opt 12(3):034034

45. Ganser H, Urban W, Brown AM (2003) The sensitive detection of NO by Faraday modulation spectroscopy with a quantum cascade laser. Mol Phys 101(4–5):545–550

46. Lewicki R, Doty JH, Curl RF et al (2009) Ultrasensitive detection of nitric oxide at 5.33 mu m by using external cavity quantum cascade laser-based Faraday rotation spectroscopy. Proc Natl Acad Sci U S A 106(31): 12587–12592

47. Kluczynski P, Lundqvist S, Westberg J et al (2011) Faraday rotation spectrometer with sub-second response time for detection of nitric oxide using a cw DFB quantum cascade laser at 5.33 mu m. Appl Phys B 103(2):451–459

48. Ganser H, Horstjann M, Suschek CV et al (2004) Online monitoring of biogenic nitric oxide with a QC laser-based Faraday modulation technique. Appl Phys B 78(3–4):513–517

Chapter 11

Electrochemical Detection of Nitric Oxide in Plant Cell Suspensions

Sophie Griveau, Angélique Besson-Bard, Fethi Bedioui, and David Wendehenne

Abstract

Nitric oxide is a hydrophobic radical acting as a physiological mediator in plants. Because of its unique properties, the detection of NO in plant tissues and cell suspensions remains a challenge. For this purpose, several techniques are used, each having certain advantages and limitations such as interferences with other species, questionable sensitivity, and/or selectivity or *ex situ* measurement. Here we describe a very attractive approach for tracking NO in plant cell suspensions using a NO-sensitive homemade platinum/iridium-based electrochemical microsensor. This method constitutes the absolute real-time proof of the production of free NO in physiological conditions.

Key words Nitric oxide, Plant cell suspensions, Electrochemical detection, Homemade electrode

1 Introduction

Nitric oxide (NO) is a diatomic hydrophobic gas recognized as a key messenger in vertebrate cardiovascular and nervous systems [1–3]. NO is also a major component of the immune system as a part of the antimicrobial armory and through regulatory functions [4, 5]. In plants, the finding that NO is a component of immunity [6, 7] initiated a new area of research and publications related to its functions in plant physiology run into thousands. Since it formally entered the scene, a myriad of plant processes were shown to involve NO, including the regulation of seed dormancy, root development, iron nutrition, flowering, stomatal closure, and adaptive responses to both biotic and abiotic stresses [8–10]. How NO mediates these processes has also received a great deal of attention in the past 15 years. Increasing bodies of evidence indicate that it fulfills the criteria of a signaling molecule. Accordingly, cross talks operating between NO and classical second messengers, reactive oxygen species (ROS), phytohormones, and protein kinase

Kapuganti Jagadis Gupta (ed.), *Plant Nitric Oxide: Methods and Protocols*, Methods in Molecular Biology, vol. 1424,
DOI 10.1007/978-1-4939-3600-7_11, © Springer Science+Business Media New York 2016

cascades were reported [10–14]. A large number of genes regulated through NO-dependent pathways and encoding proteins related to diverse cellular and physiological processes were also identified [15]. Further supporting a signaling role for NO, S-nitrosylation, and tyrosine nitration, two major posttranslational protein modifications in mammals have been shown to control a wide array of plant protein functions and emerge as fundamental mechanisms in phytohormone and stress signaling [16–20].

The possibility that plants produce NO was raised for at least 30 years ago [21, 22]. However, despite the intense interest in NO research, the sources of NO as well as the mechanisms underlying its synthesis in plants remain unclear [23]. Broadly convergent outcomes are supportive for the involvement of nitrite (NO_2^-) as a major substrate for NO synthesis through nonenzymatical as well as enzymatical processes [24–26]. In particular, nitrate reductase (NR) was shown to catalyze the reduction of NO_2^- to NO both *in vitro* and *in vivo*. For instance, NR accounts for NO synthesis in abscisic acid-induced stomatal closure [27] and in plants undergoing immune responses [28, 29]. Besides NO_2^-, studies identified polyamines [30, 31], hydroxylamines [32], and L-arginine [33] as putative substrates for NO synthesis. The finding that an L-arginine-dependent route for NO synthesis occurs in plant cells opened the possibility that they do possess a nitric oxide synthase (NOS)-like enzyme. NOS is the main enzyme responsible for NO synthesis in animals. This enzyme converts L-arginine to L-citrulline and NO radical through a series of two consecutive oxidations. NOS-like activities were detected in protein extracts from plant tissues, cell suspensions, and organelles such as peroxisomes [34]. However, efforts to identify a plant NOS have failed and the plant genomes sequenced so far do not contain genes showing partial identity to animal NOS [35, 36]. Furthermore doubts regarding the specificity of the measurement of NOS activities in plant cell extracts were underlined [37].

The difficulties encountered in identifying NO sources are also related to the analytical assays for monitoring NO. Because of its unique chemical properties, measurement of NO is technically problematic [38, 39] and its detection and quantification (as well as those of its derivatives) remain a main challenge faced by scientists, whatever the type of organisms considered. Indeed, NO is produced in a broad range of concentrations ranging from picomolar to micromolar and displays a short half-life estimated as being in the order of seconds in biological systems. Furthermore, NO is produced in subcellular compartments, diffuses, and is particularly reactive and rapidly scavenged by various cellular compounds such as other free radicals, oxygen, thiols, or metalloproteins [40]. Finally, NO signals have complex temporal and spatial arrangements, complicating its detection [41]. Thus, the analytic tools chosen for NO detection depend on the biological model and inherent features of NO production and functions. For instance, besides the sensitivity and selectivity criteria, methods for analyzing NO production must have adequate intracellular access to faithfully reflect intracellular conditions. As a consequence of these difficulties, inconsistent findings on the involvement of NO in several biological processes were reported [38].

The specificity of the methods dedicated to NO measurement has been the subject of numerous debates between plant biologists (reviewed by ref. 42). Several analytical techniques were and are still used. These methods include electron paramagnetic resonance spectroscopy (EPR), fluorometry using NO-sensitive probes, gas-phase chemiluminescence, oxyhemoglobin and Griess assays, laser photoacoustics, commercial electrodes, and CrO_3-based conversion of NO to nitrogen dioxide (NO_2) [42, 43]. Each of these methods has certain advantages and limitations including interferences with other species, questionable sensitivity and/or selectivity, or *ex situ* measurement. Furthermore, most of them are indirect as relying on the measurement of secondary species such as dinitrogen trioxide (N_2O_3), $NO_2^\bullet$, NO_2^-, nitrate (NO_3^-), or NO-adducts. Recently, new classes of fluorophores enabling a direct detection of NO such as the CuFL dye consisting in a fluorescein-based ligand (FL) complexed with Cu(II) [44] have been successfully applied [26, 45]. Their specificity for NO in plant cells still remains to be firmly established. To overcome the limitations of individual methodologies, it is now recommended to detect NO production with at least two methods [42] and to evaluate this production in the presence of pharmacological compounds scavenging/inhibiting NO synthesis or in mutants impaired in NO synthesis. Furthermore, investigating reaction products of NO such as peroxynitrite ($ONOO^-$) or molecular targets such as S-nitrosylated proteins could strengthen the demonstration of the occurrence of NO in a particular physiological process.

In addition to the methods listed above, strategies based on the use of ultramicroelectrodes deserve special attention. In animals, electrochemical detection systems are now recognized as being the only reliable method allowing a direct, real-time, label-free *in vivo* detection of NO [46, 47]. The electrochemical detection of NO is performed by inducing its electrochemical oxidation at the electrode *via* a three-electron oxidation mechanism [48–50]:

$$NO \rightarrow NO^+ + e^- \qquad (1)$$

$$NO^+ + OH^- \rightarrow HNO_2 \qquad (2)$$

$$HNO_2 + H_2O \rightarrow NO_3^- + 2e^- + 3H^+ \qquad (3)$$

The reality is that surface electrode modification is needed to make the electrode material selective for the electrochemical oxidation of NO which occurs on conventional electrode materials at high potential values, in the range of 0.7–1 V vs. saturated calomel electrode (SCE) in aqueous solutions at pH 7.4. Such high operating potential values systematically lead to concurrent oxidation of many analytes present in solution such as ascorbic acid, hydrogen peroxide (H_2O_2), and NO_2^-, thus creating interfering signals. Therefore, the design of modified electrode surfaces using organized thin layers is very attractive and provides the ideal strategy. In the general case, the chemical modification of electrode surfaces with polyelectrolytes and metal complex-based polymer films [51] has expanded the scope of application of such designed electrodes and provided a lot of options for their use in various experimental conditions. In addition to their electrocatalytic applications [52], such electrodes showed a great promise for electroanalysis [53]. As far as this aspect is concerned, substantial improvements in selectivity, sensitivity, and reproducibility can be achieved. The aim of the use of electrocatalytic material incorporated at the electrode surface is to shift the detection potential of NO toward lower values, thus improving the selectivity of detection.

To enhance the selectivity of NO electrochemical sensor, polymeric membranes are deposited onto conventional electrode surface such as platinum, gold, glassy carbon, or carbon fiber or disk electrode surfaces [54–56]. An elegant method of electrode modification is the use of electropolymerization of suitable monomers. Progress in this area was strongly connected to the design of new electroactive polymeric systems and to the success in forming thin, insoluble, stable, reproducible, and adherent films on electrode surfaces. The major additional advantage of the electropolymerization design is the ability to coat very small (microelectrodes) or irregularly shaped electrode surfaces. In the case where nonconducting polymer films are required, the electropolymerization process also offers the ability to control the film thickness, *via* the amount of charge passed, by self-regulation, since they would grow thick enough to become insulators. Thus, the use of multilayered coatings is principally aimed at ensuring the impermeability to undesired analytes by acting as efficient physical barriers. The transport properties of the various considered molecules, solvent, and supporting electrolyte across the coatings, which are affected amongst others by the film morphology, permeability, and thermal treatment, have to be well understood to better evaluate the performances of the sensors. Also, the sensitivity is crucial since the expected amount of released NO is very small (submicromolar range) so that adequate NO detection methods should have a low limit of detection. The sensitivity of the sensing materials should be analyzed in terms of hydrophobicity of the coatings. Indeed, it is now known that the pronounced hydrophobic character of NO is undoubtedly central to its physiological function, permitting the small molecule to pass freely across cell boundaries and hydrophobic membranes. Thus, the lower the degree of hydrophobicity of the microsensor coating, the less favorable is the preferential partitioning of NO from the aqueous solution to the membranes and the less sensitive is the micro electrochemical sensor.

The design and fabrication of these electrochemical sensors are now reaching a high level of sophistication and offer attractive characteristics including, amongst others, a good selectivity toward NO, an efficient discrimination against interacting species, and a good sensitivity down the nanomolar range. For example, we demonstrated the efficiency of an ultramicroelectrode-based approach for investigating NO production in plant cell suspensions elicited by cryptogein, a 10 kDa protein produced by the oomycete *Phytophthora cryptogea* [57]. NO was detected within minutes in the extracellular medium and its overall concentration was estimated as being in the order of 300 nM. These data provided clear evidence that this elicitor triggers NO production and confirms initial studies in which the occurrence of NO in cryptogein signaling was based on fluorescence assays [58, 59]. Subsequent work using CuFL dye and CrO_3-based conversion of NO to NO_2 further strengthens this finding [45, 60]. However, despite the unquestionable advantages provided by electrochemical NO sensors, this method remains underestimated by plant biologists.

This chapter describes the principle of the electrochemical detection of NO and details the technical procedures for designing this sensor and measuring NO production in the extracellular medium of plant cell suspensions.

2   Materials

2.1   Electrode Components and Equipment for NO Detection

The preparation of the NO sensor requires a Pt-Ir alloy wire (Pt 90 %-Ir 10 %, diameter 125 μm, insulated with a Teflon layer, Advent Research Materials, England) and a Ag wire for the fabrication of the pseudo-reference electrode (Ag/AgCl). The Pt-Ir wire will be chemically modified (*see* Subheading 3.1) and acts as the working electrode and the Ag/AgCl electrode as both the pseudo-reference and counter electrodes.

The preparation of Ag/AgCl electrode is obtained by applying 1.5 V between Ag and Pt wires in 3 M KCl solution for 20 min. The Ag/AgCl wire is then thoroughly rinsed with water.

Electrochemical experiments are performed using a potentiostat (Quadstat, eDAQ Pty Ltd, Australia) adapted to low-current measurements and connected to a computer. The data are collected using dedicated eChart software (eDAQ Pty Ltd, Australia), giving the current as a function of elapsed time.

2.2 Chemicals for NO Sensor Preparation	1. Nickel(II) phthalocyanine-tetrasulfonic acid tetrasodium salt (NiTSPc) 2 mM in 0.1 M NaOH aqueous solution (referred as NiTSPc solution).

2. Small volume of an ethanolic solution of Nafion perfluorinated solution (5 wt% in lower aliphatic alcohols and water) (referred as Nafion ethanolic solution).

3. *o*-Phenylenediamine (o-PD) 5 mM in 0.1 M phosphate buffer solution (referred as o-PD solution).

2.3 Cell Suspensions

The example of tobacco cell suspensions is given. The culture conditions might differ according to the plant species, the original tissues, and the labs. Prepare all solutions using ultrapure water (prepared by purifying deionized water to attain a sensitivity of 18 MΩ cm at 25 °C). All basic salts and chemicals are purchased from Sigma-Aldrich (Saint-Louis, USA).

1. Tobacco cell suspensions (*Nicotiana tabacum* L. cv *Xanthi*) are cultivated in a sterile Chandler's medium [61] on a rotary shaker (150 rpm) and under continuous light (photon flux rate 30–40 μmol/m^2/s). The culture room is maintained at 25 °C. Typically, 100 mL of cell suspensions are cultured in 250 mL glass Erlenmeyer flacks and are sub-cultured every 7 days (*see* **Note 1**).

2. H10 buffer: 175 mM mannitol, 0.5 mM CaCl$_2$, 0.5 mM K$_2$SO$_4$, 10 mM HEPES, pH 6.0. Store at 4 °C (*see* **Note 2**) and equilibrate it at 25 °C before use.

3. Cryptogein: Cryptogein is purified from the culture medium of *Phytophthora cryptogea* according to [62] and dissolved in water. Cryptogein is stored at –20 °C as a 100 μM stock solution in small aliquots to minimize freeze/thaw cycles and defrosted at room temperature before use.

4. Plastic 12-well cell culture plates (Sigma-Aldrich, Saint-Louis, USA).

5. 125 mL Sintered glass filter plate, pore size 40–100 μm (porosity 2).

3   Methods

3.1 Principle of the Electrochemical Detection of NO and Designing of the NO Sensor

NO detection is performed by inducing its electrochemical oxidation at a chemically modified Pt-Ir electrode to confer the sensitivity and the selectivity to the electrode surface toward NO. The surface of the Pt-Ir electrode is modified in three steps to obtain the NO amperometric sensor:

1. Place the Pt-Ir and Ag/AgCl wires inside NiTSPc solution and connect them to a potentiostat. Then apply a constant poten-

tial of 1.2 V for 10 min to form a homogeneous layer of NiTSPc at the electrode surface [63]. Then, rinse the electrode with water.

2. Place the modified electrode for 15 s in the Nafion ethanolic solution and let to dry at ambient temperature for 10 min. Repeat this twice.

3. Place the modified wire and Ag/AgCl wire inside the o-PD solution and connect it to a potentiostat. Then apply a constant potential of 0.9 V for 20 min to form an electropolymerized film of poly-(o-PD) at the electrode surface [47]. Finally, rinse the wire gently with water.

   Eventually, the obtained chemically modified Pt-Ir wire will act as the amperometric NO sensor.

3.2 Preparation of the Tobacco Cell Suspensions

1. Twenty hours before the assay, 10 mL of 7-day cultured cells are sub-cultured in 100 mL of sterile Chandler's medium and cultured for 24 h using the conditions described in the material section. After this step, all the manipulations are performed at room temperature without requiring sterile conditions.

2. Gently filter the 100 mL fresh cultured cells on a sintered glass filter connected to vacuum in order to remove the Chandler's medium (*see* **Note 3**).

3. Keep the filtered cells on the sintered glass filter connected to vacuum and perform three gentle washes using 50 mL of H10 buffer equilibrated at 25 °C.

4. Transfer 4 mL of washed cells in a well of a 12-well cell culture plate. We recommend setting up the cell concentration at 0.1 g of fresh weight per mL of H10 buffer (*see* **Note 4**).

5. Place the 12-well cell culture plate on a magnetic agitator and equilibrate the cells in the H10 buffer for 2 h under continuous light and shacking using a mini barrel (60 rpm).

3.3 NO Measurement

A general view of the system is presented in Fig. 1a.

1. Introduce both electrodes (NO sensor and Ag/AgCl wire) into a well of the 12-well cell culture plate. Position them just above the cells (Fig. 1b).

2. Connect both electrodes to a potentiostat.

3. Record the current while applying a constant potential of +0.6 V vs. Ag/AgCl. An amperogram showing an increase of the current which occurs after the addition of cryptogein in tobacco suspension cells is presented in Fig. 2. An equivalent volume of water is added to control cells. This treatment does not trigger an increase of current (data not shown).

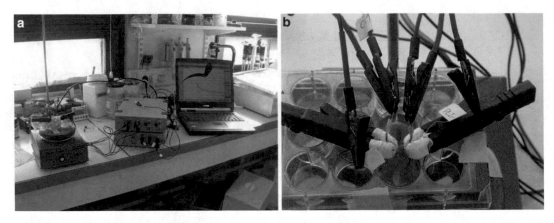

Fig. 1 Pictures of the experimental device. (**a**) General view of the experimental device with, from *left to right*, the plastic 12-well cell culture plate containing the suspension cells and the electrodes put on a magnetic agitator, the potentiostat, and the computer. (**b**) Close-up on a well of the cell culture plate containing the suspension cells in which are introduced the NO sensor and the Ag/AgCl wire. Note that two electrode systems are used simultaneously in the well

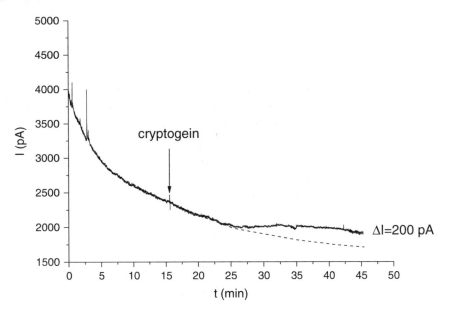

Fig. 2 Amperometric curve illustrating the increase of extracellular NO concentration in tobacco cell suspensions treated by cryptogein (100 nM). Operating potential, +0.6 V. The data are representative of six experiments

4   Notes

1. To sub-culture the cell suspension, 10 mL of 7-day cultured cells are sub-cultured in 100 mL of sterile Chandler's medium.

2. The H10 medium has been developed to measure early cellular events triggered by cryptogein in tobacco cells: alkalization of

the extracellular medium, ion fluxes across the plasma membrane, and production of ROS (*see* ref. 64 as example). The pH has been fixed at 6 in order to follow the alkalization process but also to optimize cryptogein binding to its binding sites on the plasma membrane [65]. Therefore, the composition of this medium might differ according to the type of treatment applied to the cell culture and according to the type of cells.

3. This step requires a particular attention as mechanical stress triggers NO and ROS production in plant cells which might interfere with the assay.

4. This concentration might vary according to the type of cell culture. Tobacco cells are highly responsive to cryptogein. With less responsive cells, a higher concentration might be required.

Acknowledgement

This work was supported by La Région de Bourgogne PARI AGRALE 8 project, the University of Burgundy Bonus Qualité Recherche project, and the ANR PIANO (A.B-B. and D.W.).

References

1. Foster MW, Stamler JS (2004) New insights into protein S-nitrosylation. Mitochondria as a model system. J Biol Chem 279: 25891–25897

2. Schmidt HH, Walter U (1994) NO at work. Cell 78:919–925

3. Mustafa AK, Gadalla MM, Snyder SH (2009) Signaling by gasotransmitters. Sci Signal 2:re2

4. Hernansanz-Agustin P, Izquierdo-Alvarez A, Garcia-Ortiz A, Ibiza S, Serrador JM, Martinez-Ruiz A (2013) Nitrosothiols in the immune system: signaling and protection. Antioxid Redox Signal 18:288–308

5. Bogdan C (2001) Nitric oxide and the immune response. Nat Immunol 2:907–916

6. Delledonne M, Xia Y, Dixon RA, Lamb C (1998) Nitric oxide functions as a signal in plant disease resistance. Nature 394:585–588

7. Durner J, Wendehenne D, Klessig DF (1998) Defense gene induction in tobacco by nitric oxide, cyclic GMP, and cyclic ADP-ribose. Proc Natl Acad Sci U S A 95:10328–10333

8. Besson-Bard A, Pugin A, Wendehenne D (2008) New insights into nitric oxide signaling in plants. Annu Rev Plant Biol 59:21–39

9. Wendehenne D, Gao QM, Kachroo A, Kachroo P (2014) Free radical-mediated systemic immunity in plants. Curr Opin Plant Biol 20:127–134

10. Yu M, Lamattina L, Spoel SH, Loake GJ (2014) Nitric oxide function in plant biology: a redox cue in deconvolution. New Phytol 202:1142–1156

11. Leitner M, Vandelle E, Gaupels F, Bellin D, Delledonne M (2009) NO signals in the haze: nitric oxide signalling in plant defence. Curr Opin Plant Biol 12:451–458

12. Gaupels F, Kuruthukulangarakoola GT, Durner J (2011) Upstream and downstream signals of nitric oxide in pathogen defence. Curr Opin Plant Biol 14:707–714

13. Mur LA, Prats E, Pierre S, Hall MA, Hebelstrup KH (2013) Integrating nitric oxide into salicylic acid and jasmonic acid/ ethylene plant defense pathways. Front Plant Sci 4:215

14. Jeandroz S, Lamotte O, Astier J, Rasul S, Trapet P, Besson-Bard A, Bourque S, Nicolas-Frances V, Ma W, Berkowitz GA, Wendehenne D (2013) There's more to the picture than meets the eye: nitric oxide cross talk with Ca2+ signaling. Plant Physiol 163:459–470

15. Besson-Bard A, Astier J, Rasul S, Wawer I, Dubreuil-Maurizi C, Jeandroz S, Wendehenne D (2009) Current view of nitric oxide-responsive genes in plants. Plant Sci 177:302–309

16. Astier J, Kulik A, Koen E, Besson-Bard A, Bourque S, Jeandroz S, Lamotte O, Wendehenne D (2012) Protein S-nitrosylation: what's going on in plants? Free Radic Biol Med 53:1101–1110

17. Lamotte O, Bertoldo JB, Besson-Bard A, Rosnoblet C, Aime S, Hichami S, Terenzi H, Wendehenne D (2015) Protein S-nitrosylation: specificity and identification strategies in plants. Front Chem 2:114

18. Skelly MJ, Loake GJ (2013) Synthesis of redox-active molecules and their signaling functions during the expression of plant disease resistance. Antioxid Redox Signal 19:990–997

19. Corpas FJ, Chaki M, Leterrier M, Barroso JB (2009) Protein tyrosine nitration: a new challenge in plants. Plant Signal Behav 4:920–923

20. Vandelle E, Delledonne M (2011) Peroxynitrite formation and function in plants. Plant Sci 181:534–539

21. Nelson RS, Ryan SA, Harper JE (1983) Soybean mutants lacking constitutive nitrate reductase activity: I. Selection and initial plant characterization. Plant Physiol 72:503–509

22. Klepper L (1990) Comparison between NO(x) evolution mechanisms of wild-type and nr(1) mutant soybean leaves. Plant Physiol 93:26–32

23. Gupta KJ, Fernie AR, Kaiser WM, van Dongen JT (2011) On the origins of nitric oxide. Trends Plant Sci 16:160–168

24. Yamasaki H, Sakihama Y (2000) Simultaneous production of nitric oxide and peroxynitrite by plant nitrate reductase: in vitro evidence for the NR-dependent formation of active nitrogen species. FEBS Lett 468:89–92

25. Rockel P, Strube F, Rockel A, Wildt J, Kaiser WM (2002) Regulation of nitric oxide (NO) production by plant nitrate reductase in vivo and in vitro. J Exp Bot 53:103–110

26. Horchani F, Prevot M, Boscari A, Evangelisti E, Meilhoc E, Bruand C, Raymond P, Boncompagni E, Aschi-Smiti S, Puppo A, Brouquisse R (2011) Both plant and bacterial nitrate reductases contribute to nitric oxide production in Medicago truncatula nitrogen-fixing nodules. Plant Physiol 155:1023–1036

27. Desikan R, Griffiths R, Hancock J, Neill S (2002) A new role for an old enzyme: nitrate reductase-mediated nitric oxide generation is required for abscisic acid-induced stomatal closure in Arabidopsis thaliana. Proc Natl Acad Sci U S A 99:16314–16318

28. Chen J, Vandelle E, Bellin D, Delledonne M (2014) Detection and function of nitric oxide during the hypersensitive response in Arabidopsis thaliana: where there's a will there's a way. Nitric Oxide 43:81–88

29. Rasul S, Dubreuil-Maurizi C, Lamotte O, Koen E, Poinssot B, Alcaraz G, Wendehenne D, Jeandroz S (2012) Nitric oxide production mediates oligogalacturonide-triggered immunity and resistance to Botrytis cinerea in Arabidopsis thaliana. Plant Cell Environ 35:1483–1499

30. Tun NN, Santa-Catarina C, Begum T, Silveira V, Handro W, Floh EI, Scherer GF (2006) Polyamines induce rapid biosynthesis of nitric oxide (NO) in Arabidopsis thaliana seedlings. Plant Cell Physiol 47:346–354

31. Yang B, Wu J, Gao F, Wang J, Su G (2014) Polyamine-induced nitric oxide generation and its potential requirement for peroxide in suspension cells of soybean cotyledon node callus. Plant Physiol Biochem 79:41–47

32. Rumer S, Gupta KJ, Kaiser WM (2009) Plant cells oxidize hydroxylamines to NO. J Exp Bot 60:2065–2072

33. Crawford NM, Guo F-Q (2005) New insights into nitric oxide metabolism and regulatory functions. Trends Plant Sci 10:195–200

34. Corpas FJ, Palma JM, Del Río LA, Barroso JB (2009) Evidence supporting the existence of l-arginine-dependent nitric oxide synthase activity in plants. New Phytol 184:9–14

35. Butt Y, Lum J, Lo S (2003) Proteomic identification of plant proteins probed by mammalian nitric oxide synthase antibodies. Planta 216:762–771

36. Frohlich A, Durner J (2011) The hunt for plant nitric oxide synthase (NOS): is one really needed? Plant Sci 181:401–404

37. Tischner R, Galli M, Heimer YM, Bielefeld S, Okamoto M, Mack A, Crawford NM (2007) Interference with the citrulline-based nitric oxide synthase assay by argininosuccinate lyase activity in Arabidopsis extracts. FEBS J 274:4238–4245

38. Csonka C, Páli T, Bencsik P, Görbe A, Ferdinandy P, Csont T (2015) Measurement of NO in biological samples. Br J Pharmacol 172:1620–1632

39. Hetrick EM, Schoenfisch MH (2009) Analytical chemistry of nitric oxide. Annu Rev Anal Chem 2:409–433

40. Stamler JS, Singel DJ, Loscalzo J (1992) Biochemistry of nitric oxide and its redox-activated forms. Science 258:1898–1902

41. Hess DT, Matsumoto A, Kim SO, Marshall HE, Stamler JS (2005) Protein S-nitrosylation:

purview and parameters. Nat Rev Mol Cell Biol 6:150–166

42. Mur LA, Mandon J, Cristescu SM, Harren FJ, Prats E (2011) Methods of nitric oxide detection in plants: a commentary. Plant Sci 181:509–519

43. Vandelle E, Delledonne M (2008) Methods for nitric oxide detection during plant-pathogen interactions. Methods Enzymol 437:575–594

44. Lim MH, Xu D, Lippard SJ (2006) Visualization of nitric oxide in living cells by a copper-based fluorescent probe. Nat Chem Biol 2:375–380

45. Kulik A, Noirot E, Grandperret V, Bourque S, Fromentin J, Salloignon P, Truntzer C, Dobrowolska G, Simon-Plas F, Wendehenne D (2015) Interplays between nitric oxide and reactive oxygen species in cryptogein signalling. Plant Cell Environ 38:331–348

46. Griveau S, Bedioui F (2013) Overview of significant examples of electrochemical sensor arrays designed for detection of nitric oxide and relevant species in a biological environment. Anal Bioanal Chem 405:3475–3488

47. Griveau S, Dumézy C, Séguin J, Chabot GG, Scherman D, Bedioui F (2006) In vivo electrochemical detection of nitric oxide in tumor-bearing mice. Anal Chem 79:1030–1033

48. Ciszewski A, Milczarek G (2003) Electrochemical detection of nitric oxide using polymer modified electrodes. Talanta 61:11–26

49. Privett BJ, Shin JH, Schoenfisch MH (2010) Electrochemical nitric oxide sensors for physiological measurements. Chem Soc Rev 39:1925–1935

50. Bedioui F, Villeneuve N (2003) Electrochemical nitric oxide sensors for biological samples – principle, selected examples and applications. Electroanalysis 15:5–18

51. Murray RW (1992) Molecular design of electrode surfaces. Wiley, New York, NY

52. Gilmartin MA, Hart JP (1995) Sensing with chemically and biologically modified carbon electrodes. A review. Analyst 120:1029–1045

53. Bakker E, Telting-Diaz M (2002) Electrochemical sensors. Anal Chem 74:2781–2800

54. Shibuki K (1990) An electrochemical microprobe for detecting nitric oxide release in brain tissue. Neurosci Res 9:69–76

55. Bedioui F, Quinton D, Griveau S, Nyokong T (2010) Designing molecular materials and strategies for the electrochemical detection of nitric oxide, superoxide and peroxynitrite in biological systems. Phys Chem Chem Phys 12:9976–9988

56. Bedioui F, Griveau S (2013) Electrochemical detection of nitric oxide: assessment of twenty years of strategies. Electroanalysis 25:587–600

57. Besson-Bard A, Griveau S, Bedioui F, Wendehenne D (2008) Real-time electrochemical detection of extracellular nitric oxide in tobacco cells exposed to cryptogein, an elicitor of defence responses. J Exp Bot 59:3407–3414

58. Lamotte O, Gould K, Lecourieux D, Sequeira-Legrand A, Lebrun-Garcia A, Durner J, Pugin A, Wendehenne D (2004) Analysis of nitric oxide signaling functions in tobacco cells challenged by the elicitor cryptogein. Plant Physiol 135:516–529

59. Foissner I, Wendehenne D, Langebartels C, Durner J (2000) In vivo imaging of an elicitor-induced nitric oxide burst in tobacco. Plant J 23:817–824

60. Vitecek J, Reinohl V, Jones RL (2008) Measuring NO production by plant tissues and suspension cultured cells. Mol Plant 1:270–284

61. Chandler MT, Marsac NTD, Kouchkovsky YD (1972) Photosynthetic growth of tobacco cells in liquid suspension. Can J Bot 50:2265–2270

62. Bourque S, Dutartre A, Hammoudi V, Blanc S, Dahan J, Jeandroz S, Pichereaux C, Rossignol M, Wendehenne D (2011) Type-2 histone deacetylases as new regulators of elicitor-induced cell death in plants. New Phytol 192:127–139

63. Trévin S, Bedioui F, Devynck J (1996) New electropolymerized nickel porphyrin films. Application to the detection of nitric oxide in aqueous solution. J Electroanal Chem 408:261–265

64. Wendehenne D, Lamotte O, Frachisse JM, Barbier-Brygoo H, Pugin A (2002) Nitrate efflux is an essential component of the cryptogein signaling pathway leading to defense responses and hypersensitive cell death in tobacco. Plant Cell 14:1937–1951

65. Wendehenne D, Binet MN, Blein JP, Ricci P, Pugin A (1995) Evidence for specific, high-affinity binding sites for a proteinaceous elicitor in tobacco plasma membrane. FEBS Lett 374:203–207

Chapter 12

Quantification and Localization of *S*-Nitrosothiols (SNOs) in Higher Plants

Juan B. Barroso, Raquel Valderrama, Alfonso Carreras, Mounira Chaki,
Juan C. Begara-Morales, Beatriz Sánchez-Calvo,
and Francisco J. Corpas

Abstract

S-nitrosothiols (SNOs) are a family of molecules produced by the reaction of nitric oxide (NO) with –SH thiol groups present in the cysteine residues of proteins and peptides caused by a posttranslational modification (PTM) known as *S*-nitrosylation (strictly speaking *S*-nitrosation) that can affect the cellular function of proteins. These molecules are a relatively more stable form of NO and consequently can act as a major intracellular NO reservoir and, in some cases, as a long-distance NO signal. Additionally, SNOs can be transferred between small peptides and protein thiol groups through *S*-transnitrosylation mechanisms. Thus, detection and cellular localization of SNOs in plant cells can be useful tools to determine how these molecules are modulated under physiological and adverse conditions and to determine their importance as a mechanism for regulating different biochemical pathways. Using a highly sensitive chemiluminescence ozone technique and a specific fluorescence probe (Alexa Fluor 488 Hg-link phenylmercury), the methods described in this chapter enable us to determine SNOs in an nM range as well as their cellular distribution in the tissues of different plant species.

Key words Chemiluminescence, Confocal laser scanning microscope, Fluorescent probes, Nitric oxide, Reactive nitrogen species, *S*-nitrosothiol

1 Introduction

Nitric oxide (NO) is a gaseous free radical that affects a wide variety of physiological and pathological aspects of higher plants [1–5]. Nitric oxide can interact with many molecules that affect the function of target molecules in many cases. For example, the chemical reaction between NO and the superoxide radical ($O_2^{\bullet-}$) generates peroxynitrite ($ONOO^-$) which is the most powerful oxidant that mediates PTM through nitration reactions [6, 7]. On the other hand, the interaction of NO with –SH thiol groups present in the cysteine residues of peptides and proteins generates a group of molecules called *S*-nitrosothiols (SNOs) [8–10]. This NO, which

Kapuganti Jagadis Gupta (ed.), *Plant Nitric Oxide: Methods and Protocols*, Methods in Molecular Biology, vol. 1424,
DOI 10.1007/978-1-4939-3600-7_12, © Springer Science+Business Media New York 2016

binds with thiols to form S-nitrosothiols, is biologically reversible
[11]. In addition, its molecular family is regarded as one of the
most important mechanisms for transducing signals mediated by
NO in biological systems [11–13]. This is explained by the fact that
a process of S-nitrosylation—strictly speaking S-nitrosation [11]—is
capable of regulating the function of target proteins [14–18].

In general, one group of SNOs contains designated high-
molecular-mass SNOs produced by NO binding to sulfhydryl
(–SH) groups present in specific cysteine residues of proteins.
Another group is composed of low-molecular-mass SNOs, the
most important one being S-nitrosoglutathione (GSNO) which is
generated by NO's S-nitrosylation reaction with the thiol tripep-
tide, γ-glutamyl cysteinyl glycine (glutathione, GSH), one of the
major low-molecular-weight soluble antioxidants in plant cells [19,
20]. This group includes other molecules such as S-nitrosocysteine
(CySNO) and S-nitrosocysteinylglycine (GlyCySNO) [21], which
have been studied to a lesser degree in the field of plant research.

The detection and quantification of SNOs in biological systems
and specifically in higher plants is a major challenge as the presence of
artifacts can quite easily produce misleading information (reviewed in
ref. [22]). This chapter provides two complementary approaches which
can be useful for studying the content and distribution of S-nitrosothiols
in different plant tissues and species under physiological and stress
conditions and confirms the involvement of these molecules in the
response and signaling mechanism in different situations [23–28].

2 Materials

2.1 Specific Equipment

1. Nitric oxide analyzer (NOA): Sievers, model 280i, Sievers
   Instruments.

2. Preparative centrifuge (Sorvall RC-5 and rotors).

3. Vibratome (Leica Microsystems, Germany).

4. Confocal laser scanning microscope (CLSM) system (Leica
   TCS SL, Leica Microsystems, Wetzlar, Germany).

2.2 Reagents

1. Nitrite (NO_2^-) standard stock solution: 100 mM.

2. N-ethylmaleimide (NEM) stock solution: 100 mM.

3. Sulfanilamide stock solution: 100 mM.

4. Mercury (II) chloride ($HgCl_2$): $HgCl_2$ (molecular weight = 271.5)
   in 10 ml ultrapure water: 13.7 mg.

5. Reaction solution (iodine/triiodide mixture) for the NOA
   purge vessel: 4.5 ml of glacial acetic acid and 500 μl aqueous
   mixture containing 450 mM potassium iodide (KI) and
   100 mM iodine (I_2). Dissolve 108 mg KI in 1 ml ultrapure

water. Add 38 mg I_2 and mix at room temperature. Mix 1 ml of this solution with 9 ml glacial acetic.

6. Alexa Fluor 488 Hg-link phenylmercury (cat. no. H30462, Molecular Probes, Eugene, OR, USA).

2.3 Solutions

1. The extraction medium (1:5; w/v) was made up of 100 mM Tris–HCl buffer (pH 7.5) containing 5 % (w/v) sucrose, 7 % (w/v) PVPP, 0.05 % (v/v) Triton X-100, 0.1 mM EDTA, 15 mM DTT, 1 mM PMSF, 100 µM DTPA, and a protease inhibitor cocktail.

3 Methods

3.1 Ozone Chemiluminescence Detection of Total and Low-Molecular-Mass SNOs

The ozone-based chemiluminescence technique has become a reliable method to detect and quantify NO and other related molecules including SNOs as it is highly sensitive, having in some cases a concentration range of nM. The method used in plants to quantify total SNOs is based on the technique described by Valderrama et al. [23] with some modifications [26]. The detection of SNOs is based on the reductive decomposition of nitroso species by an iodine/triiodide mixture to release NO, which is then measured with the aid of gas-phase chemiluminescence in reaction with ozone [29]. Unlike other nitroso species such as nitrosamines and nitrosyl hemes, SNOs are sensitive to reductants and mercury-induced decomposition.

3.1.1 Plant Extracts

1. During all procedures, the samples were kept at 4 °C and protected from the light.

2. Plant samples were grounded using a mortar and pestle in liquid nitrogen.

3. The resulting coarse powder was transferred to 1/5 (w/v) extraction buffer, which can change depending on the plant species and tissue involved. For example, for pea leaf samples, we successfully used a medium (1:3; w/v) containing 100 mM Tris–HCl, pH 7.6, 1.5 mM dithiothreitol (DTT), 5 % sucrose (w/v), 0.005 % Triton X-100 (v/v), and 100 µM diethylene-tetraminepentaacetic acid (DTPA).

4. The extract was then centrifuged at $3000 \times g$ for 10 min at 4 °C.

5. The supernatant was used for SNO determination. In the case of sunflower hypocotyls, the extraction medium is more complex as this tissue can oxidize very rapidly (brown color).

6. The supernatant was used to determine total SNOs and the level of low-molecular-mass SNOs was determined in a filtered supernatant fraction that had previously passed through a 10 kDa cutoff membrane.

Note: For the chemiluminescence detection of total SNOs, the whole procedure needs to be performed under a red safety light in order to protect SNOs from light-dependent decomposition.

3.1.2 Treatment of Plant Samples to Block Free Thiols and to Eliminate Nitrite Content

Add 10 mM NEM to the plant samples (for each 1.5 ml sample, add 167 μl of 100 mM NEM) and incubate for 15 min at 4 °C in order to block free thiols.

For each sample, three 1.5 ml aliquots were prepared:

1. Aliquot A: Inject directly into the nitric oxide analyzer (NOA) with its corresponding blank.
2. Aliquot B: Add 10 mM sulfanilamide and incubate for 15 min at 4 °C. This will eliminate nitrite from the samples.
3. Aliquot C: Add 10 mM sulfanilamide and 7.3 mM $HgCl_2$. Incubate for 30 min at room temperature. This will eliminate nitrite and SNOs, respectively.

Calculations

Nitrite concentration = Aliquot A – Aliquot B

S-nitrosothiol concentration = Aliquot B – Aliquot C

Other nitroso species = Aliquot C

Set up the NOA and samples.

1. Prepare samples and treatments (aliquots A, B, and C).
2. Prepare purge chamber of the NOA containing 5 ml iodine/triiodide mixture at 95 °C under a nitrogen stream.
3. Prepare nitrite standard curve between 1 nM and 100 mM prepared from a standard 100 mM nitrite stock solution.
4. Set up the NOA.
   (a) Set up a water bath at 70 °C and open the NOA refrigeration system.
   (b) Open the N_2 and O_2 system and maintain a constant flux (0.5 bar).
   (c) Add 5 ml reaction solution (potassium iodide/iodine mix) to the purge vessel. It is recommended to change this solution every 4–5 injections.
5. Inject samples (40 μl or more depending on the signal obtained) into the NOA purge chamber.

Figure 1a represents total SNO content in pea leaf samples from plants exposed to different abiotic stress situations [24]. Figure 1b shows total SNO content in the hypocotyls of two sunflower cultivars after infection by *Plasmopara halstedii* [25].

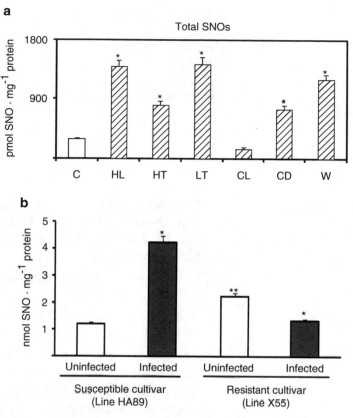

Fig. 1 Analysis of total S-nitrosothiols (SNOs) determined by ozone chemilumi-nescence in (**a**) pea leaves of plants exposed to different abiotic stress conditions and (**b**) sunflower hypocotyls after infection by *Plasmopara halstedii*. (**a**) Leaf extracts from pea plants subjected to different abiotic stress conditions. *C* control, *HL* high light intensity, *HT* high temperature, *LT* low temperature, *CL* continuous light, *CD* continuous dark, *W* wounding. Results are means ± SEM of samples from at least three different experiments. *Asterisks* indicate that the increase in SNOs was statistically significant ($P > 0.005$; $n \geq 3$) as compared with control plants (C). Reproduced with permission from Corpas et al. [24]. (**b**) Hypocotyl extracts from two sunflower cultivars (HA89 and X55, susceptible and resistant cultivars, respectively), uninfected and after infection by *P. halstedii*. Results are the mean of four different experiments ± SEM. *Differences between uninfected and infected values were significant at $P < 0.05$. **Differences between control (uninfected) values were significant at $P < 0.05$. Reproduced with permission from Chaki et al. [25]

3.2 Localization of SNOs with the Aid of a Confocal Laser Scanning Microscope

In plant tissues, SNOs can also be detected using the fluorescent reagent Alexa Fluor 488 Hg-link phenylmercury. The method used to study the cell distribution of total SNOs in plants is based on the technique described by Valderrama et al. [23] with some modifications [30]. This dye, which is a new generation of fluorescent probes, has certain advantages over other fluorescent dyes; for example, it is highly fluorescent over a broad pH range (pH 4–10), has a higher level of photostability, allows more time for image capture, has good water solubility, and is less affected by autofluorescence from proteins and other biomolecules. Thus, Alexa Fluor 488 Hg-Link phenyl mercury can be used for direct *S*-nitrosothiol detection in plant tissues when samples are preincubated with *N*-ethylmaleimide (NEM) to block free sulfhydryl groups. Under these conditions, the fluorescent Hg-Link reagent labels R-SNO-modified proteins and peptides according to the following equation:

$$R^1 \!-\!\!\bigcirc\!\!-\!Hg^+ \; + \; R\text{-}SNO \; \longrightarrow \; R^1 \!-\!\!\bigcirc\!\!-\!HgS\text{-}R$$

Hg-link phenylmercury   Thiolate

3.2.1 Procedure

1. Plant tissue is cut into segments of approximately 25 mm² and is incubated at 25 °C for 2 h in darkness, with 100 μM (DTPA) plus 10 mM NEM prepared in ethanol, which blocks free sulfhydryl groups.

2. Each segment is washed three times in 10 mM Tris–HCl buffer, pH 7.4, for 15 min.

3. Segments are incubated with 10 μM Alexa Fluor 488 Hg-Link phenyl mercury for 1 h at 25 °C in darkness.

4. Segments are washed again three times in the previous buffer.

5. Plant segments are embedded in order to obtain appropriate sections to be observed under a confocal laser scanning microscope in a 15 % acrylamide solution prepared in 0.01 M phosphate-buffered saline (PBS) including 0.3 % TEMED for 4 h.

6. Place each plant segment in inclusion containers (1.5×0.9×0.5 mm; Sorvall Instruments) and add 0.5 ml of a fresh 15 % acrylamide stock solution and 50 μl of 2 % persulfate ammonium (PSA) to be polymerized. Plant segments must be quickly oriented after PSA is added, and the containers are covered with parafilm in order to improve polymerization.

7. After the acrylamide block containing the plant segments is polymerized, latter are removed.

8. Sections measuring 80–100 μm thick are cut, as indicated by the Vibratome scale, under 10 mM PBS. The sections are then soaked in a glycerol:PBS mixture containing azide (1:1, v/v) and mounted in the same medium on a slide for examination with the aid of a confocal laser scanning microscope (CLSM) system (Leica TCS SL) using standard filters for Alexa Fluor 488 green fluorescence (excitation 495 nm; emission 519 nm).

All procedures must be performed under a red safety light.

3.2.2 Additional Controls

SNOs are capable of releasing NO in a controlled manner into the living system either spontaneously or through interaction with various biological reductants such as glutathione and ascorbate. In addition, the reduced metal ion (e.g., Cu⁺) breaks SNOs down more rapidly than the oxidized metal ion (e.g., Cu²⁺), indicating that reducing agents such as glutathione and ascorbate can cause the breakdown of S-nitrosothiol by the chemical reduction of contaminating transition-metal ions [31–33]. To confirm the effect of these reductants, as a control, the detection of SNOs in plant samples can be carried out in the presence of reductants. Although the procedure used is similar to the previous one, in this case, the plant tissue segments are preincubated in a reductant solution containing 1 mM ascorbate, 10 μM CuCl, and 200 μM cPTIO in 10 mM phosphate-buffered saline (PBS) at room temperature for 1 h. Afterwards, each sample is washed three times for 15 min (two washes with PBS and one with 10 mM Tris–HCl buffer, pH 7.4). The procedure then proceeds from **steps 1** to **8** as described above.

Figure 2 shows an example of the cellular localization of SNOs in hypocotyl sections of sunflower seedlings growing under optimal

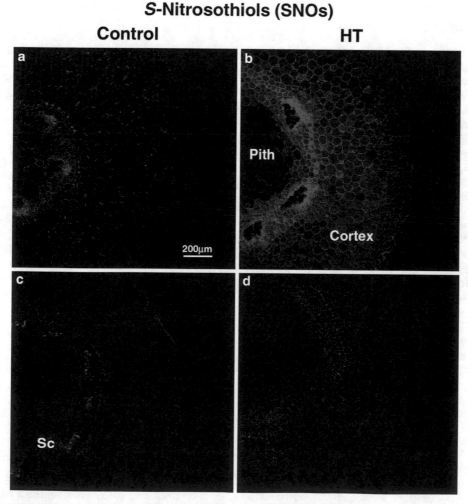

Fig. 2 Images illustrating the CLSM detection of SNOs in cross sections of hypocotyls in sunflower seedlings from control and subjected to high temperature (HT) in the absence of reductants (A and B) and in the presence of reductants (D and E). Bright green fluorescence corresponds to detection of SNO with 10 μM Alexa Fluor 488 Hg-link. *Sc* sclereids. Reproduced with permission from Chaki et al. [26]

conditions (Fig. 2a) and exposed to high temperature (HT) where green fluorescence indicates the presence of SNOs [26]. It can thus be observed how HT stress provokes a significant increase in SNO content in cortex cells and sclereids. On the other hand, Fig. 2c, d shows the additional internal controls where the hypocotyl section from sunflower seedlings growing under both optimal and stress conditions was preincubated with a reductant solution which causes a significant reduction in SNO content.

Acknowledgments

This work was supported by an ERDF-cofinanced grant from the Ministry of Science and Innovation (BIO2012-33904) and Junta de Andalucía (groups BIO192 and BIO286).

References

1. Lamattina L, García-Mata C, Graziano M et al (2003) Nitric oxide: the versatility of an extensive signal molecule. Annu Rev Plant Biol 54:109–136

2. Neill SJ, Desikan R, Clarke A et al (2003) Nitric oxide signalling in plants. New Phytol 159:11–35

3. Shapiro AD (2005) Nitric oxide signaling in plants. Vitam Horm 72:339–398

4. Corpas FJ, Leterrier M, Valderrama R et al (2011) Nitric oxide imbalance provokes a nitrosative response in plants under abiotic stress. Plant Sci 181:604–611

5. Besson-Bard A, Pugin A, Wendehenne D (2008) New insights into nitric oxide signaling in plants. Annu Rev Plant Biol 59:21–39

6. Corpas FJ, del Río LA, Barroso JB (2007) Need of biomarkers of nitrosative stress in plants. Trends Plant Sci 12:436–438

7. Radi R (2013) Peroxynitrite, a stealthy biological oxidant. J Biol Chem 288:26464–26472

8. Stamler JS, Singel DJ, Loscalzo J (1992) Biochemistry of nitric oxide and its redox-activated forms. Science 258:1898–1902

9. Jourd'heuil D, Jourd'heuil FL, Lowery AM et al (2005) Detection of nitrosothiols and other nitroso species in vitro and in cells. Methods Enzymol 396:118–131

10. Broniowska KA, Hogg N (2012) The chemical biology of S-nitrosothiols. Antioxid Redox Signal 17:969–980

11. Smith BC, Marletta MA (2012) Mechanisms of S-nitrosothiol formation and selectivity in nitric oxide signaling. Curr Opin Chem Biol 16:498–506

12. Corpas FJ, del Río LA, Barroso JB (2008) Post-translational modifications mediated by reactive nitrogen species: nitrosative stress responses or components of signal transduction pathways? Plant Signal Behav 3:301–303

13. Gould N, Doulias PT, Tenopoulou M et al (2013) Regulation of protein function and signaling by reversible cysteine S-nitrosylation. J Biol Chem 288:26473–26479

14. Lindermayr C, Saalbach G, Durner J (2005) Proteomic identification of S-nitrosylated proteins in Arabidopsis. Plant Physiol 137:921–930

15. Yun BW, Feechan A, Yin M et al (2011) S-nitrosylation of NADPH oxidase regulates cell death in plant immunity. Nature 478:264–268

16. Astier J, Kulik A, Koen E et al (2012) Protein S-nitrosylation: what's going on in plants? Free Radic Biol Med 53:1101–1110

17. Begara-Morales JC, Chaki M, Sánchez-Calvo B et al (2013) Protein tyrosine nitration in pea roots during development and senescence. J Exp Bot 64:1121–1134

18. Corpas FJ, Begara-Morales JC, Sánchez-Calvo B et al (2015) Nitration and S-nitrosylation: two post-translational modifications (PTMs) mediated by reactive nitrogen species (RNS) which participate in signalling processes of plant cells. In: Gupta KJ, Igamberdiev AU (eds) Reactive oxygen and nitrogen species signaling and communication in plants, vol 23, Signaling and communication in plants. Springer, New York. doi:10.1007/978-3-319-10079-1_13.ISBN978-3-319-10078-4

19. Noctor G, Mhamdi A, Chaouch S et al (2012) Glutathione in plants: an integrated overview. Plant Cell Environ 35:454–484

20. Corpas FJ, Alché JD, Barroso JB (2013) Current overview of S-nitrosoglutathione (GSNO) in higher plants. Front Plant Sci 4:126

21. Doulias PT, Raju K, Greene JL et al (2013) Mass spectrometry-based identification of S-nitrosocysteine in vivo using organic mercury assisted enrichment. Methods 62:165–170

22. Diers AR, Keszler A, Hogg N (2014) Detection of S-nitrosothiols. Biochim Biophys Acta 1840: 892–900

23. Valderrama R, Corpas FJ, Carreras A et al (2007) Nitrosative stress in plants. FEBS Lett 581:453–461

24. Corpas FJ, Chaki M, Fernández-Ocaña A et al (2008) Metabolism of reactive nitrogen species in pea plants under abiotic stress conditions. Plant Cell Physiol 49:1711–1722

25. Chaki M, Fernández-Ocaña AM, Valderrama R et al (2009) Involvement of reactive nitrogen and oxygen species (RNS and ROS) in sunflower-mildew interaction. Plant Cell Physiol 50:265–279

26. Chaki M, Valderrama R, Fernández-Ocaña AM et al (2011) High temperature triggers the metabolism of S-nitrosothiols in sunflower mediating a process of nitrosative stress which provokes the inhibition of ferredoxin-NADP reductase by tyrosine nitration. Plant Cell Environ 34:1803–1818

27. Chaki M, Valderrama R, Fernández-Ocaña AM et al (2011) Mechanical wounding induces a nitrosative stress by down-regulation of GSNO reductase and an increase in S-nitrosothiols in sunflower (Helianthus annuus) seedlings. J Exp Bot 62:1803–1813

28. Begara-Morales JC, Sánchez-Calvo B, Chaki M et al (2014) Dual regulation of cytosolic ascorbate peroxidase (APX) by tyrosine nitration and S-nitrosylation. J Exp Bot 65:527–538

29. Feelisch M, Rassaf T, Mnaimneh S et al (2002) Concomitant S-, N-, and heme-nitros(yl)ation in biological tissues and fluids: implications for the fate of NO in vivo. FASEB J 16:1775–1785

30. Corpas FJ, Carreras A, Esteban FJ et al (2008) Localization of S-nitrosothiols and assay of nitric oxide synthase and S-nitrosoglutathione reductase activity in plants. Methods Enzymol 437:561–574

31. Singh RJ, Hogg N, Joseph J, Kalyanaraman B et al (1996) Mechanism of nitric oxide release from S-nitrosothiols. J Biol Chem 271:18596–18603

32. Williams DLH (1996) The mechanism of nitric oxide formation from S-nitrosothiols. Chem Commun 10:1085–1090

33. Smith JN, Dasgupta TP (2000) Kinetics and mechanism of the decomposition of S-nitrosoglutathione by L-ascorbic acid and copper ions in aqueous solution to produce nitric oxide. Nitric Oxide 4:57–66

Chapter 13

Analysis of the Expression and Activity of Nitric Oxide Synthase from Marine Photosynthetic Microorganisms

Noelia Foresi, Natalia Correa-Aragunde, Jerome Santolini, and Lorenzo Lamattina

Abstract

Nitric oxide (NO) functions as a signaling molecule in many biological processes in species belonging to all kingdoms of life. In animal cells, NO is synthesized primarily by NO synthase (NOS), an enzyme that catalyze the NADPH-dependent oxidation of l-arginine to NO and l-citrulline. Three NOS isoforms have been identified, the constitutive neuronal NOS (nNOS) and endothelial NOS (eNOS) and one inducible (iNOS). Plant NO synthesis is complex and is a matter of ongoing investigation and debate. Despite evidence of an Arg-dependent pathway for NO synthesis in plants, no plant NOS homologs to animal forms have been identified to date. In plants, there is also evidence for a nitrate-dependent mechanism of NO synthesis, catalyzed by cytosolic nitrate reductase. The existence of a NOS enzyme in the plant kingdom, from the tiny single-celled green alga *Ostreococcus tauri* was reported in 2010. *O. tauri* shares a common ancestor with higher plants and is considered to be part of an early diverging class within the green plant lineage.

In this chapter we describe detailed protocols to study the expression and characterization of the enzymatic activity of NOS from *O. tauri*. The most used methods for the characterization of a canonical NOS are the analysis of spectral properties of the oxyferrous complex in the heme domain, the oxyhemoglobin (oxyHb) and citrulline assays and the NADPH oxidation for in vitro analysis of its activity or the use of fluorescent probes and Griess assay for in vivo NO determination. We further discuss the advantages and drawbacks of each method. Finally, we remark factors associated to the measurement of NOS activity in photosynthetic organisms that can generate misunderstandings in the interpretation of results.

Key words Nitric oxide, Nitric oxide synthase, Oxyhemoglobin assay, Citrulline detection, Griess assay, DAF-FM diacetate

1 Introduction

Nitric oxide synthases (NOSs, EC 1.14.13.39) are present in insects, mollusks, parasites, fungi, slime molds, bacteria and recent genome sequencing revealed that NOS proteins exist in photosynthetic organisms [1–5]. Their amino acid sequences and activities are similar to the mammalian NOSs, suggesting that the mammalian gene came from lower species through evolution [6]. The prokaryotic forms have been considered to be the precursors of animals

Kapuganti Jagadis Gupta (ed.), *Plant Nitric Oxide: Methods and Protocols*, Methods in Molecular Biology, vol. 1424,
DOI 10.1007/978-1-4939-3600-7_13, © Springer Science+Business Media New York 2016

NOS. The last were originated during evolution by the fusion of a bacterial oxygenase domain to a dedicated reductase domain. The bacterial enzymes generally have only the oxygenase domain highly homologous to the mammalian counterpart. However, the NOS from the bacterium *Sorangium cellulosum* contains a reductase domaintogether the oxygenase, with a different arrangement of the domains between bacterial and animal NOS enzymes, suggest independent events in prokaryotic and eukaryotic lineages [4, 7].

The mammalian NOSs catalyze the oxidation of l-Arginine to l-citrulline and NO, with *N*-hydroxy-l-Arg formed as an enzyme-bound intermediate as shown in [A].

H$_2$N—NH$_2$ 1 NADPH →   H$_2$N—N–OH   0.5 NADPH →   O—NH$_2$ •N=O

O$_2$ H$_2$O   O$_2$ H$_2$O

L-Arginine   N^N-Hydroxy-L-Arginine   L-Citrulline   Nitric Oxide

All mammalian NOSs are bi-domain proteins comprising an N-terminal oxygenase domain (NOSoxy) that binds protoporphyrin IX (heme), 6R-tetrahydrobiopterin (BH$_4$), and Arg, and a C-terminal flavoprotein domain (NOSred), linked together by a calmodulin (CaM) binding sequence. NOS flavoprotein domains are similar to NADPH-cytochrome P450 reductase and related electron transfer flavoproteins, and function to provide NADPH-derived electrons to the ferric heme for O$_2$ activation during NO synthesis [8].

The identification of the NOS from the green algae *O. tauri* represents the first NOS characterized in the plant kingdom [5]. By sequence homology analysis, more NOS sequences can be found in other photosynthetic organisms. *O. lucimarinus*, another species of the Ostreococcus genus, also contains a NOS gene, the expression of which was validated by EST analysis [5]. In *Bathycoccus prasinos* genome there is a sequence with 75 % (accession number XP_007510938) and in *Thalassiosira oceanica* genome (accession number EJK55330) a sequence with 56 % similarity with the *Ostreococcus tauri* NOS (OtNOS), indicating that novel NOS proteins will be discovered in the plant kingdom in the future. The big challenge remaining undeciphered is to identify and characterize the NOS activity in higher plants.

As sated before, OtNOS is the unique NOS enzyme characterized in photosynthetic organisms [5]. OtNOS has a 42 % similarity to human NOS reaching to 45–49 % similarity to invertebrate

NOS. OtNOS contains the NOSoxy and NOSred domains joined by a CaM binding domain. Despite the high similarity, some differences could be noted in the structure of the OtNOS with respect to animal NOS. CaM plays a critical role in activating NOS, since it triggers the electron transfer from flavin to heme. OtNOS activity behaves like an intermediate between eNOS/nNOS and iNOS (endothelial, neuronal and inducible NOS respectively) isoforms since in the absence of Ca^{2+}-CaM, OtNOS retains almost 70 % of activity. Furthermore, OtNOS lacks of the autoregulatory control element (ACE) [5], indicating that it is close to the mammalian iNOS isoform. The ACE impedes CaM binding and enzymatic activation in constitutive NOSs (eNOS and nNOS). The increase in Ca^{2+} concentration triggers the binding of Ca^{2+}-CaM in constitutive NOSs by displacing the ACE [9]. BH_4 cofactor is essential for NO production in animals since the absence of BH_4 uncouples the reaction leading to NADPH oxidation and superoxide formation [10]. Ostreococcus genome has been completely sequenced [11] and it lacks the genes encoding for the enzymes that synthesize BH_4, suggesting that OtNOS may bind another cofactor for catalytic activity. Thus, it is potentially a useful model system to study gene evolution and cellular processes in photosynthetic eukaryotes.

A good approach for the characterization of a canonical NOS enzyme is the recombinant expression of the NOS protein in *Escherichia coli*. This strategy allows to express and purify the protein for analyzing its spectral characteristics and in vitro activity. Furthermore, in *E. coli* expressing the recombinant NOS protein it can be measured NO production and nitrite levels, an indirect method used to quantify NO concentration in vivo. Increased NO and nitrite levels were detected in *E. coli* expressing NOS recombinant enzyme from mammalian, bacterial, and photosynthetic organisms [5, 6]. However, the detection of the NOS activity in vivo in the original organism still remains essential. Foresi et al. [5] reported the functional characterization of the NOS enzyme from *O. tauri* by heterologous expression in *E. coli*. Bacterium carrying the NOS gene displayed enhanced NO production and cell viability. In *O. tauri*, OtNOS protein was detected throughout its life cycle participating in light-regulated response of the algae. In this protocol, the methodology used to study expression and the activity of NOS in the green algae will be described and discussed in detail.

2 Biological Materials

Cloning, expression and purification of heterologous expression of recombinant NOS in *E. coli* is a useful approach to study the NOS activity in vitro and in vivo.

2.1 Cloning of Recombinant NOS

1. Synthesizing, sequencing and cloning the full length NOS sequence or the oxy domain (CaM domain, inclusive) of NOS into pET expression system vector (*see* **Note 1**).

2. Transform competent BL21 (DE3) protease-deficient *E. coli* cells with pET-NOS (*see* **Note 2**).

2.2 Recombinant NOS Expression

2.2.1 NOSoxi Expression

1. Inoculate Erlenmeyer flasks containing 0.5 L of modified Terrific Broth (20 g of yeast extract, 10 g of bacto tryptone, 2.65 g of KH_2PO_4, 4.33 g of Na_2HPO_4, and 4 mL of glycerol) and ampicillin (125 µg/mL) with 500 µL of culture pET15b-NOS and grown overnight and then shaking at 150 rpm at 37 °C.

2. Induce the recombinant protein expression at OD_{600} between 1 and 1.2 by the addition of 1 mM IPTG (*see* **Notes 3** and **4**). Add the heme precursor 8-aminolevulinic acid at final concentrations of 500 µM.

3. Harvest the cells after 72 h at 20 °C (*see* **Note 4**) of induction and resuspend in 250 mL of Lysis buffer (100 mM Tris–HCl, pH 7.4, 150 mM NaCl, 1 mg/mL lysozyme, 10 % [v/v] glycerol, 1 mM PMSF, 1 mg/L leupeptin, and 0.5 mg/L pepstatin, 5 mg/L aprotinin, 10 mM l-Arg, 10 µM BH_4 in ascorbic acid, 50 U/mL of DNAse). Lysis is achieved by cell disruption using a Cell disruptor Brand or a French press at 0.6/0.8 kbar.

2.2.2 Full NOS Expression

1. Inoculate Fernbach flasks containing 1 L of modified Terrific Broth and kanamycin (50 mg/mL) with 1 mL of culture pET-NOS and grown over night shaking at 190 rpm at 37 °C.

2. Induce the recombinant protein expression at OD_{600} 0.4 by the addition of 0.5 mM IPTG (*see* **Notes 3** and **4**). Add the heme and flavin precursors 8-aminolevulinic acid and riboflavin at final concentrations of 450 and 3 µM, respectively.

3. Harvest the cells after 24 h (*see* **Note 4**) of induction and resuspend in 30 mL of buffer (100 mM Tris–HCl, pH 7.4, 1 mM EDTA, 1 mM DTT, 10 % [v/v] glycerol, 1 mM PMSF, 5 mg/mL leupeptin, and 5 mg/mL pepstatin) per liter of initial culture and lysed by pulsed sonication (six cycles of 20 s).

4. Analyze the best conditions for NOS expression testing different concentrations of IPTG and time of induction. Recombinant NOS is then studied by SDS-PAGE and Western blot analysis using a specific anti-NOS antibody or anti-H is when this tag is added (*see* **Note 5**).

2.3 NOS Purification

1. For OtNOS oxi domain, remove cell debris by centrifugation ($20,000 \times g$). The supernatant is applied on a NTA-Ni^{2+} column (20 mL), previously equilibrated in MCAC buffer (100 mM Tris–HCl, pH 7.4, 10 % glycerol, 150 mM NaCl, 1 mM PMSF, 10 mM l-Arg, 10 µM BH_4 in ascorbic acid). Column is extensively washed (8 column volumes) with a MCAC + 40

mM imidazole buffer. Elution is achieved by the addition of a MCAC + 300 mM imidazole buffer.

2. For full NOS purification remove cell debris by centrifugation, and apply the supernatant to an ADP-agarose 4B column (1 mL) equilibrated in buffer B (50 mM Tris–HCl, pH 7.4, 0.1 mM EDTA, 0.1 mM DTT, 10 % glycerol, and 100 mM NaCl) (*see* **Note 6**). Wash the column with 10 column volumes of buffer B and finally with buffer B and 500 mM NaCl. Elute the protein with buffer B, 500 mM NaCl, and 25 mM 2′-AMP (*see* **Note 7**).

3   Methods

Carry out all the procedures at 4 °C temperature, NOS activity is quickly lost at room temperature.

3.1 Spectral Properties of the Oxygenase Domain of NOS

1. Recondition the oxygenase domain in a KPi, pH 7.4 10 % [v/v] glycerol and 150 mM NaCl Buffer by three cycles of concentration/dilution.

2. Analyze the UV–visible absorption spectra (Fig. 1 main) are then recorded for the native enzyme (around 20 μM) and upon successive additions of l-Arginine (10 mM final) and BH$_4$ (40 μM final). Fig. 1 Inset displays the characteristic spectrum of the Fe^{II}CO complex obtained upon addition of sodium dithionite and CO flushing.

3.2 Determination of NOS Activity In Vitro

3.2.1 Spectro-photometric Method

1. Reduce the hemoglobin to oxyhemoglobin with sodium dithionite (*see* **Note 8**) Determine the oxyhemoglobin concentration in the solution using a molar extinction coefficient of 131 mM^{-1} cm^{-1} at 415 nm.

2. Prepare a 500 μL-mix reaction containing 20 mM oxyhemoglobin, 7.5 mM HEPES-NaOH, pH 7.5, 5 mM DTT, 100 μM l-arginine, 1 μM NADPH, 10 mM CaCl$_2$, 10 μM CaM, 100 μM BH$_4$, and 100 U/mL catalase.

3. Initiate the reaction by the addition of 0.5 μM purified NOS protein. Monitor the NO-dependent conversion of oxyhemoglobin to methemoglobin on a spectrophotometer by scanning between 380 and 450 nm. Use an extinction coefficient of 100 mM^{-1} cm^{-1} to quantify NO production as the difference between the peak at 401 nm and the valley at 420 nm.

3.2.2 NADPH Oxidation

1. Prepare 500 μL volume containing 0.5 μM NOS, 50 mM Tris–HCl, pH 7.6, 5 mM DTT, 100 μM l-Arg, 10 mM CaCl$_2$, 10 μM CaM, 100 μM BH$_4$, and 100 U/mL catalase and start the reaction with 1 μM NADPH (*see* **Note 9**).

2. Monitor the rate of decrease in absorbance at 340 nm for 10 min at 25 °C using a spectrophotometer. An extinction

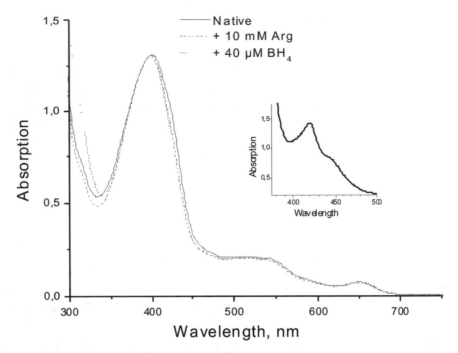

Fig. 1 UV–visible absorption spectra of the oxygenase domain of OtNOS. *Main panel*: Native OtNOS (*solid line*) shows spectroscopic fingerprints charecteristic of mammalian NOSs: a Soret maximum around 402 nm (indicating a mixture of high spin and low spin species), an α/β region with two bands around 515 and 545 nm and a charge transfer band at 650 nm. Addition of the substrate L-arginine (*dashed line*) and the cofactor BH$_4$ (*dotted line*) shifts the mixture to a fully high spin species (I_{max} at 398 nm). *Inset*: P450 spectra of OtNOS. The Fe^{II}CO complex of OtNOS shows a mixture between P450 and P420 conformations, indicating a certain liability of the proximal ligation

coefficient of 6.22 mM^{-1} cm^{-1} at 340 nm is used to calculate NADPH oxidation.

3.2.3 Citrulline Detection

1. Perform the enzymatic reaction at 25 °C in 50 mM Tris–HCl, pH 7.4, containing 50 μM l-Arg, 1 μCi [3H] l-Arginine monohydrochloride (40–70 Ci/mmol; Perkin-Elmer), 100 μM NADPH, 10 μM FAD, 2 mM CaCl$_2$, 1 μg CaM, and 100 μM BH$_4$ in a volume of 40 μL (*see* **Note 10**).

2. Initiate enzymatic reactions by adding 0.5 μM NOS and terminated after 5–30 min by the addition of 400 μL of ice-cold 20 mM sodium acetate, pH 5.5, containing 1 μM l-citrulline, 2 mM EDTA, and 0.2 mM EGTA (stop buffer).

3. Apply the sample to columns containing 1 mL of Dowex AG50W-X8, Na$^+$ form (Bio-Rad; 100–200 mesh), pre-equilibrated in stop buffer.

4. Elute l-citrulline with 2 mL of distilled water. Aliquots of 0.5 mL of eluate are dissolved in 10 mL of scintillation liquid consisting of 0.35 % (w/v) PPO (2,5-diphenyloxazole), 0.03 % (w/v) POPPOP (1,4-bis[2-(5-phenyloxazolyl)]benzene),

1 % (w/v) naphthalene, 2.3 % (v/v) ethanol, 3.85 % (v/v) diox-ane, and 3.85 % (v/v) toluene. Radioactivity is measured in a liquid scintillation counter.

For a correct citrulline assay interpretation a thin layer chroma-tography is necessary (*see* **Note 10**). The citrulline production by OtNOS was evidenced by TLC (Fig. 2a, b).

1. Prepare the reaction: 50 mM Tris–HCl (pH 7.4) containing 50 µM l-Arg, 1 µCi [^{3}H]l-arginine monohydrochloride, 100 µM NADPH, 10 µM FAD, 2 mM CaCl$_2$, 1 µg CaM, and 10 µM BH$_4$ in a final volume of 40 µL.

2. Start the reaction by the addition of 0.5 µM OtNOS or H$_2$O (Control).

3. After 30 min at 25 °C, stop the enzymatic reaction with 40 µL of 20 mM sodium acetate pH 5.5, containing 2 mM EDTA and 0.2 mM EGTA, and load onto a Dowex AG 50 W-X8 resin column.

4. Elute the citrulline by centrifugation and separate it by TLC.

5. Run standard solutions contained 0.2 µmol of l-Arg, 0.2 µmol l-citrulline and a mix containing 0.1 µmol of l-Arg plus 0.1 µmol of l-citrulline.

6. Separate the amino acids on Silica-60 TLC plates (Merck) employing chloroform, methanol and ammonium hydroxide (2:3:2, v/v/v) as the mobile phase.

7. Stain the amino acids with ninhydrin solution (0.1 %, w/v, ninhydrin in ethanol–acetic acid, 5:1, v/v).

8. Remove the amino acids spots from TLC plates and determine the radioactivity using a liquid scintillation counter.

3.3 In Vivo Assay Method to Detect NOS Activity

3.3.1 Griess Assay

The in vivo NOS activity can be analyzed in the bacteria expressing the recombinant protein.

1. Determination of NO formation using the Griess reagent in *E. coli* (*see* **Note 11**). For a nitrite measurement inoculate Fernbach flasks with 500 µL of culture containing 50 mL of LB medium and shake at 190 rpm at 37 °C. Induce recombi-nant protein expression by the addition of IPTG (*see* **Note 4**).

2. Add the heme and flavin precursors 8-aminolevulinic acid and riboflavin, to 450 and 1 µM, respectively, and the substrate l-Arg to 1 mM, all of them final concentrations.

3. The cells are harvested 5 h after the induction and the pellet is solubilized in 100 mM phosphate buffer pH 7.5 and lysated by pulsed sonication (six cycles of 20 s each).

4. Remove cell debris by centrifugation, and use the supernatant for Griess assay. Fifty microliters of sample is placed in a 96-well micro assay plate.

a

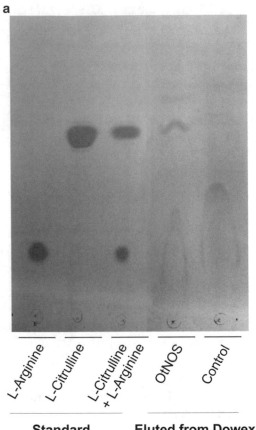

b

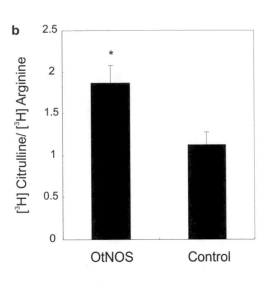

©2010 by American Society of Plant Biologists

Fig. 2 Detection of L-citrulline as a product of OtNOS enzymatic activity. (**a**) Thin layer chromatography (TLC) of the products of reactions. Retension factors (Rf) were determined (L-arginine, 0.21; L-citrulline, 0.59). (**b**) Amino acids were removed from TLC plates and radioactivity was determined using a liquid scintillation counter. Values are expressed as the ratio of [³H]L-citrulline to [³H]L-arginine. *Error bars* denote SE ($n=4$) and *asterisk* indicates a statistically significant difference (*t* test, $p < 0.05$). Foresi et al., 2010. www.plantcell.org. Copyright American Society of Plant Biologists

5. Add 50 µL of sulfanamide in 5 % [v/v] phosphoric acid; incubate for 10 min at 20 °C.

6. Add 50 µL of 1 % (w/v) *N*-(1-naphthyl) ethylenediamine HCl (NED) and incubate for 10 additional min in the dark.

7. Read the absorbance at 550 nm in a microplate reader. A nitrite standard calibration curve (0–100 µM) is performed for nitrite concentration estimation.

3.3.2 Fluorometric Detection of NO Production

1. Add DAF-FM diacetate(10 µM, Molecular Probes) to the culture medium and incubate in the dark for 20 min prior measurement (*see* **Note 12**).

2. Measure NO fluorescence intensity (excitation 495 nm; emission 515 nm) using a fluorescence plate reader or visualize NO dependent green fluorescence under an inverted fluorescence microscope. A culture without NOS protein as negative control is essential, with the same optical density (*see* **Note 12**).

4   Notes

1. The DNA sequence can be obtained from RNA extracted from the organism and preparing cDNA with transcriptase reverse and DNA by PCR or DNA synthesized commercially from genomic sequence databases [5]. The pET expression system (Novagen) is one of the most used systems for the cloning and in vivo expression of recombinant proteins in *E. coli*. This is due to the high selectivity of the pET system's bacteriophage T7 RNA polymerase for its cognate promoter sequences, the high level of activity of the polymerase and the high translation efficiency mediated by the T7 gene *10* translation initiation signals. In the pET system, the protein coding sequence of interest is cloned downstream of the T7 promoter and gene *10* leader sequences, and then transformed into *E. coli* strains. Protein expression is achieved by IPTG induction. Due to the specificity of the T7 promoter, basal expression of cloned target genes is extremely low in strains lacking a source of T7 RNA polymerase. This phenomenon, together with high-efficiency translation, achieves expression levels in which the target protein may constitute the majority of the cellular protein after only a few hours [12].

2. The pET24b-OtNOS vector is used to transform BL21 (DE3) protease-deficient *E. coli* via electroporation [5]. The BL21 (DE3) competent cells are an all purpose strain for high-level protein expression and easy induction. This strain is deficient in the proteases ompT and lon, resulting in a suitable host for recombinant expression [13].

3. *E.coli* is the most frequently used host for production of NOS enzymes [6, 14, 15]. Only mammalian iNOS requires of CaM co-expression to generate a fully active form expressed in *E. coli* [16].

4. The IPTG concentration used for NOS expression depends on the protein and the vector used for the expression system; it can vary between 0.1 and 2 mM. The temperature and time of induction depends on the NOS protein. The optimal conditions should be tested for each protein.

5. OtNOS expression can be identified by specific anti-OtNOS antibody [5].

6. The activities of the three NOS recombinant isoforms from animals are found distributed between the soluble and particulate fractions of cells. Isoform I (nNOS from brain) and isoform II (iNOS from cytokine-induced macrophages) are mostly soluble proteins. Isoform III (eNOS) from endothelial cells is myristoylated and found predominantly in the particulate fraction. The activities of isoforms I and III are regulated

by Ca^{2+} in the nanomolar range. The activity of isoform II is Ca^{2+} independent. Similar, but not identical procedures are used to purify the different isozymes. All three isoforms are hemoproteins and require the same cofactors, NADPH (6R)-5,6,7,8-BH$_4$, flavin adenine dinucleotide (FAD), and flavin mononucleotide (FMN). Flavins and BH$_4$ are found bound to the purified enzymes in quantities that are sometimes, but not always, sufficient for full activity [17].

7. The OtNOS and eNOS recombinant proteins were purified using 2′5′-ADP Sepharose 4B and appeared as a single band of apparent molecular mass 119 and 135 kDa, respectively, on SDS/PAGE [5, 14]. 2′5′ADP Sepharose 4B media with immobilized NADPH structural analog is primarily used for purification of enzymes requiring NADPH as a cofactor.

   The addition of a specific tag (e.g., His tag) is a common strategy used to express and purify NOS proteins. NOS cDNA sequences from *Bacillus subtilis* and *Deinococcus radiodurans* were cloned with a His$_6$ tag attached to its N terminus and the proteins were over expressed in *E. coli* strain BL21 (DE3) and purified using chromatography on Ni^{2+}-nitrilotriacetic acid resin [18, 19].

8. Spectrophotometry has been widely used to measure NOS activity through the oxyhemoglobin (oxyHb) assay. NO stoichiometrically reacts with oxyHb to produce methemoglobin (metHb) [20]. The distinct optical behaviors of these species (oxyHb and metHb) make the conversion of oxyHb to metHb a simple, practical a rapid spectrophotometric determination of NOS activity [20]. Nevertheless; same precautions have to be taken when using this assay. The isolated NOS protein should be preferably not contaminated with other proteins, normally a minimal metHb formation is observed without proteins in the sample. Hence, it is necessary the control reaction performed either (a) with a specific inhibitor of NOS, (b) in absence of the cofactor nicotinamide adenine dinucleotide phosphate (NADPH) or (c) without the substrate l-Arg. This assay is commonly used for the analysis of the catalytic properties of NOS [5, 19–21].

9. NOS isoforms catalyze other leak and side reactions, such as superoxide production at the expense of NADPH. As such, this stoichiometry is not generally observed, and reflects the three electrons supplied by NADPH to form one NO. A partial uncouple reaction in the NOS activity can be detected by NADPH oxidation [22].

10. The l-citrulline assay is based on the stoichiometric production of NO and l-[^{3}H]-citrulline from l-[^{3}H]-arginine by NOS. This chemical reaction is the basis of the l-citrulline

assay, a simple and specific method that is currently used to measure NOS enzymatic activity, with control reactions (blanks) performed either (a) with a specific inhibitor of NOS, (b) in absence of the cofactor NADPH, or (c) with a protein extract that has been boiled prior to the incubation. Since the reaction can be performed with or without Ca^{2+}, it can be discriminated between Ca^{2+}-dependent and Ca^{2+}-independent NOS activities. The l-citrulline assay has been widely used to demonstrate NOS activity in a variety of cells and tissues [19, 23, 24]. However, NOS activity is relatively unstable, and the l-citrulline assay is technically inapplicable when applied with low levels of NOS protein [25].

The principle of the technique is to retain the labeled l-[^{3}H]-arginine, by the Dowex and measured in the scintillation system. There are many reports of an arginine-dependent NOS activity in higher plants. To NOS activity in plants extracts, many studies have relied on a citrulline-based assay that measures the formation of l-citrulline from l-Arg using ion exchange chromatography. However, Tischner et al. [26] reported that when such assays were used with protein extracts from *Arabidopsis*, an l-Arg-dependent NOS activity was observed, but it generated a metabolite product other than l-citrulline. Indeed, TLC analysis identified the product as argininosuccinate. The reaction was stimulated by fumarate (>500 µM), suggesting that the enzyme involved was the urea cycle enzyme argininosuccinate lyase (EC 4.3.2.1), which reversibly converts l-Arg and fumarate to argininosuccinate. These results indicate that caution is needed when using standard citrulline-based assays to measure NOS activity in plant extracts, and highlight the importance of verifying the identity of the product as l-citrulline.

11. NO has a short half-life (<10 s), which makes it difficult to detect and study. However, as NO is metabolized to nitrate and nitrite in the cell, quantization of these stable anions can be used to measure the amount of NO that was originally present in a sample [26]. The Griess reaction was reported by Johann Peter Griess in 1879 as a method of analysis of nitrite. In this method, nitrite is first treated with a diazotizing reagent (Reagent A), e.g., sulfanilamide (SA), in acidic media to form a transient diazonium salt. This intermediate is then allowed to react with a coupling Reagent B, N-naphthyl-ethylenediamine (NED), to form a stable azo compound. The intense purple color of the product allows assaying nitrite with high sensitivity and can be used to measure nitrite concentration as low as ~0.5 µM. The absorbance at 540 nm of the formed adduct is linearly proportional to the nitrite concentration in the sample [27]. Through the years, many variations

on the original reaction have been described. The most popular version seems to be the sequential method in which nitrite is mixed with SA first, followed by the addition of NED. This method seems to give highest yield of the chromophore, and therefore it is the most sensitive way to perform Griess Reaction assay [28, 29]. For a more accurate measurement of NO produced in a sample, the nitrate formed via oxidation of nitrite must also be measured. This is often accomplished by reducing nitrate to nitrite immediately prior to the addition of the Griess reagents to the initial sample [30]. This approach was employed to identify NOS activity in *E. coli* expressing a recombinant NOS, including OtNOS and NOS from *Bacillus subtilis* [5, 6]. The NOS activity in vitro can be measured by Griess reagent, however, NADPH, an essential cofactor for NOS activity, interferes with the Griess reaction, which severely limits the sensitivity of conventional assays. The addition of lactate dehydrogenase (LDH) eliminates NADPH and can increase the sensitivity of the Griess reaction.

Some considerations and recommendations for the Griess assay:

- Sulfanilamide and NED compete for nitrite in the Griess reaction [30], and thus greater sensitivity is achieved when the two components are added sequentially.
- The final pH of a sample after addition of reagent A is critical for the Griess reaction. Lower final pH results in higher absorbance at 540 nm. When the pH is lower than 1.8, the absorbance at 540 nm is stable [27]. For samples with a high buffer capacity more acid should be added to reagent A.
- In a condition where there is low level of NO production, high amount of nitrate (or nitrite) in the media will make the measurement difficult due to the high background [30], and thus it is important to know minutely nitrite and nitrate concentration in the media broth used and also to minimize as much as possible the content of these compounds from water in the solutions.

12. DAF-FM   (4-aminomethyl-5-methylamino-2′,7′-difluoro-fluorescein) and DAF-FM diacetate represent two important reagent for quantification of low concentrations of NO (~5 nM). Developed by Kojima et al. [33], these compounds are essentially nonfluorescent until they react with NO to form a fluorescent benzotriazole. DAF-FM diacetate is cell-permeable and passively diffuses across cellular membranes. Once inside cells, it is deacetylated by intracellular esterases to become DAF-FM. With excitation/emission maxima of 495/515 nm, DAF-FM can be detected by any instrument that can detect fluorescein, including flow cytometers, microscopes, fluorescent

microplate readers, and fluorometers. Probably the most successful indicator for nitric oxide has been 4,5-diaminofluorescein diacetate (DAF-2 diacetate), which was also developed by Kojima and collaborators [33, 34]. DAF-2 has been used to identify NO-production in plant cells [33]. The DAF-FM reagent has some important advantages over DAF-2. The spectra of the NO adduct of DAF-FM are independent of pH above pH 5.5. Also, the NO adduct of DAF-FM is significantly more photo-stable than that of DAF-2 which means additional time for image capture. Finally, DAF-FM is a more sensitive reagent for NO than is DAF-2 (NO detection limit for DAF-FM ~3 nM vs. ~5 nM for DAF-2) [32–34]. DAF-FM diacetate was used for NOS activity detection in *E. coli* expressing OtNOS [5]. Gusarov et al. [6] utilized the NO-specific fluorescent probe CuFL, which has only recently become available [35], to directly monitor NO production by bNOS expressed in *E. coli*. A cell culture without NOS protein as negative control is essential within a current experimental protocol.

References

1. Muller U (1997) The nitric oxide system in insects. Prog Neurobiol 51(3):363–381
2. Klessig DF, Durner J, Noad R, Navarre DA, Wendehenne D, Kumar D, Zhou JM, Shah J, Zhang S, Kachroo P, Trifa Y, Pontier D, Lam E, Silva H (2000) Nitric oxide and salicylic acid signaling in plant defense. Proc Natl Acad Sci U S A 97(16):8849–8855
3. Golderer G, Werner ER, Leitner S, Grobner P, Werner-Felmayer G (2001) Nitric oxide synthase is induced in sporulation of *Physarum polycephalum*. Genes Dev 15(10):1299–1309. doi:10.1101/gad.890501
4. Agapie T, Suseno S, Woodward JJ, Stoll S, Britt RD, Marletta MA (2009) NO formation by a catalytically self-sufficient bacterial nitric oxide synthase from *Sorangium cellulosum*. Proc Natl Acad Sci U S A 106(38):16221–16226. doi:10.1073/pnas.0908443106
5. Foresi N, Correa-Aragunde N, Parisi G, Calo G, Salerno G, Lamattina L (2010) Characterization of a nitric oxide synthase from the plant kingdom: NO generation from the green alga *Ostreococcus tauri* is light irradiance and growth phase dependent. Plant Cell 22(11):3816–3830. doi:10.1105/tpc.109.073510
6. Gusarov I, Starodubtseva M, Wang ZQ, McQuade L, Lippard SJ, Stuehr DJ, Nudler E (2008) Bacterial nitric-oxide synthases operate without a dedicated redox partner. J Biol Chem 283(19):13140–13147. doi:10.1074/jbc.M710178200

7. Andreakis N, D'Aniello S, Albalat R, Patti FP, Garcia-Fernàndez J, Procaccini G, Sordino P, Palumbo A (2011) Evolution of the nitric oxide synthase family in metazoans. Mol Biol Evol 28(1):163–179. doi:10.1093/molbev/msq179
8. Alderton WK, Cooper CE, Knowles RG (2001) Nitric oxide synthases: structure, function and inhibition. Biochem J 357(Pt 3):593–615
9. Salerno JC, Harris DE, Irizarry K, Patel B, Morales AJ, Smith SM, Martasek P, Roman LJ, Masters BS, Jones CL, Weissman BA, Lane P, Liu Q, Gross SS (1997) An autoinhibitory control element defines calcium-regulated isoforms of nitric oxide synthase. J Biol Chem 272(47):29769–29777
10. Wever RM, van Dam T, van Rijn HJ, de Groot F, Rabelink TJ (1997) Tetrahydrobiopterin regulates superoxide and nitric oxide generation by recombinant endothelial nitric oxide synthase. Biochem Biophys Res Commun 237(2):340–344. doi:10.1006/bbrc.1997.7069
11. Derelle E, Ferraz C, Rombauts S, Rouze P, Worden AZ, Robbens S, Partensky F, Degroeve S, Echeynie S, Cooke R, Saeys Y, Wuyts J, Jabbari K, Bowler C, Panaud O, Piegu B, Ball SG, Ral JP, Bouget FY, Piganeau G, De Baets B, Picard A, Delseny M, Demaille J, Van de Peer Y, Moreau H (2006) Genome analysis of the smallest free-living eukaryote *Ostreococcus tauri* unveils many unique features. Proc Natl Acad Sci U S A 103(31):11647–11652. doi:10.1073/pnas.0604795103

12. Studier FW (1991) Use of bacteriophage T7 lysozyme to improve an inducible T7 expression system. J Mol Biol 219(1):37–44

13. Sorensen HP, Mortensen KK (2005) Soluble expression of recombinant proteins in the cytoplasm of *Escherichia coli*. Microb Cell Factories 4(1):1. doi:10.1186/1475-2859-4-1

14. Martasek P, Liu Q, Liu J, Roman LJ, Gross SS, Sessa WC, Masters BS (1996) Characterization of bovine endothelial nitric oxide synthase expressed in *E. coli*. Biochem Biophys Res Commun 219(2):359–365. doi:10.1006/bbrc.1996.0238

15. Fossetta JD, Niu XD, Lunn CA, Zavodny PJ, Narula SK, Lundell D (1996) Expression of human inducible nitric oxide synthase in *Escherichia coli*. FEBS Lett 379(2):135–138

16. Wu C, Zhang J, Abu-Soud H, Ghosh DK, Stuehr DJ (1996) High-level expression of mouse inducible nitric oxide synthase in Escherichia coli requires coexpression with calmodulin. Biochem Biophys Res Commun 222(2):439–444. doi:10.1006/bbrc.1996.0763

17. Forstermann U, Closs EI, Pollock JS, Nakane M, Schwarz P, Gath I, Kleinert H (1994) Nitric oxide synthase isozymes. Characterization, purification, molecular cloning, and functions. Hypertension 23(6 Pt 2):1121–1131

18. Adak S, Aulak KS, Stuehr DJ (2002) Direct evidence for nitric oxide production by a nitric-oxide synthase-like protein from *Bacillus subtilis*. J Biol Chem 277(18):16167–16171. doi:10.1074/jbc.M201136200

19. Wang ZQ, Lawson RJ, Buddha MR, Wei CC, Crane BR, Munro AW, Stuehr DJ (2007) Bacterial flavodoxins support nitric oxide production by *Bacillus subtilis* nitric-oxide synthase. J Biol Chem 282(4):2196–2202. doi:10.1074/jbc.M608206200

20. Ghafourifar P, Asbury ML, Joshi SS, Kincaid ED (2005) Determination of mitochondrial nitric oxide synthase activity. Methods Enzymol 396:424–444. doi:10.1016/S0076-6879(05)96036-9

21. Rodriguez-Crespo I, Ortiz de Montellano PR (1996) Human endothelial nitric oxide synthase: expression in *Escherichia coli*, coexpression with calmodulin, and characterization. Arch Biochem Biophys 336(1):151–156

22. Griffith OW, Stuehr DJ (1995) Nitric oxide synthases: properties and catalytic mechanism. Annu Rev Physiol 57:707–736. doi:10.1146/annurev.ph.57.030195.003423

23. Knowles RG, Salter M (1998) Measurement of NOS activity by conversion of radiolabeled arginine to citrulline using ion-exchange separation. Methods Mol Biol 100:67–73

24. Ward TR, Mundy WR (1999) Measurement of the nitric oxide synthase activity using the citrulline assay. Methods Mol Med 22:157–162. doi:10.1385/0-89603-612-X:157

25. Combet S, Balligand JL, Lameire N, Goffin E, Devuyst O (2000) A specific method for measurement of nitric oxide synthase enzymatic activity in peritoneal biopsies. Kidney Int 57(1):332–338. doi:10.1046/j.1523-1755.2000.00839.x

26. Tischner R, Galli M, Heimer YM, Bielefeld S, Okamoto M, Mack A, Crawford NM (2007) Interference with the citrulline-based nitric oxide synthase assay by argininosuccinate lyase activity in Arabidopsis extracts. FEBS J 274(16):4238–4245. doi:10.1111/j.1742-4658.2007.05950.x

27. Xu J, Xu X, Verstraete W (2000) Adaptation of E. coli cell method for micro-scale nitrate measurement with the Griess reaction in culture media. J Microbiol Methods 41(1):23–33

28. Verdon CP, Burton BA, Prior RL (1995) Sample pretreatment with nitrate reductase and glucose-6-phosphate dehydrogenase quantitatively reduces nitrate while avoiding interference by NADP+ when the Griess reaction is used to assay for nitrite. Anal Biochem 224(2):502–508. doi:10.1006/abio.1995.1079

29. Guevara I, Iwanejko J, Dembinska-Kiec A, Pankiewicz J, Wanat A, Anna P, Golabek I, Bartus S, Malczewska-Malec M, Szczudlik A (1998) Determination of nitrite/nitrate in human biological material by the simple Griess reaction. Clin Chim Acta 274(2):177–188

30. Arita NO, Cohen MF, Tokuda G et al (2007) Fluorometric detection of nitric oxide with diaminofluoresceins (DAFs): applications and limitations for plant NO research. In: Polacco JC, Lamattina L (eds) Nitric oxide in plant growth, development and stress physiology. Springer, Berlin, pp 269–280

31. Fiddler RN (1977) Collaborative study of modified AOAC method of analysis for nitrite in meat and meat products. J Assoc Off Anal Chem 60(3):594–599

32. Foissner I, Wendehenne D, Langebartels C, Durner J (2000) In vivo imaging of an elicitor-induced nitric oxide burst in tobacco. Plant J 23(6):817–824

33. Kojima H, Nakatsubo N, Kikuchi K, Kawahara S, Kirino Y, Nagoshi H, Hirata Y, Nagano T (1998) Detection and imaging of nitric oxide with novel fluorescent indicators: diaminofluoresceins. Anal Chem 70(13):2446–2453

34. Kojima H, Urano Y, Kikuchi K, Higuchi T, Hirata Y, Nagano T (1999) Fluorescent indicators for imaging nitric oxide production. Angew Chem 38(21):3209–3212

35. Lim MH, Lippard SJ (2006) Fluorescent nitric oxide detection by copper complexes bearing anthracenyl and dansyl fluorophore ligands. Inorg Chem 45(22):8980–8989. doi:10.1021/ic0609913

Chapter 14

Identification of *S*-Nitrosothiols by the Sequential Cysteine Blocking Technique

Rafael A. Homem, Thierry Le Bihan, Manda Yu, and Gary J. Loake

Abstract

Here, we describe a procedure for the identification of *S*-nitrosothiols that has been used in our laboratory to study the roles of protein *S*-nitrosylation in the immune responses of *Arabidopsis thaliana* and other organisms. It employs a modified version of the biotin-switch technique, which we termed the sequential cysteine blocking technique, encompassing the sequential redox-blocking of recombinant proteins followed by LC–MS/MS analysis.

Key words *S*-nitrosylation, *S*-nitrosothiols, Redox-based posttranslational modification, LC–MS/MS, Iodoacetamide, Sequential cysteine blocking, Label-free quantitation

1 Introduction

Cysteine is among the least frequent amino acids incorporated into proteins in several organisms. It is characterized by the presence of a thiol moiety which is often involved in enzymatic reactions, serving as a nucleophile. In addition, the thiol moiety is subject to oxidization to give the disulfide derivative cysteine, which serves an important structural role and also functions as a site of redox sensing in many proteins. Cysteine, due to its reactivity is involved in several posttranslational modifications. The nucleophilic thiol group permits cysteine to be conjugated to other groups, for example, a farnesyl or geranyl–geranyl moiety in prenylation and a nitric oxide (NO) moiety in *S*-nitrosylation, amongst others. In Table 1, we have highlighted some of the most common natural and chemical modifications of cysteine.

S-nitrosylation is a reversible and dynamic redox-based post-translational modification (PTM) process that consists of the incorporation of a NO moiety into a protein thiol group (SH) to form an *S*-nitrosothiol group (SNO) [1]. Akin to other PTMs, *S*-nitrosylation can modulate protein activity [2], cellular compartmentalization, and DNA binding [3, 4]. Thus, *S*-nitrosylation is

Kapuganti Jagadis Gupta (ed.), *Plant Nitric Oxide: Methods and Protocols*, Methods in Molecular Biology, vol. 1424,
DOI 10.1007/978-1-4939-3600-7_14, © Springer Science+Business Media New York 2016

Table 1
List of some of the most common natural and chemical modifications of cysteine, its monoisotopic and average masses and compositions

Description	Monoisotopic mass	Average mass	Composition
S-nitrosylation	28.99016	28.998	H(−1) N O
Cysteine (SS bond)	−1.00783	−1.008	H(−1)
S-sulphenation	15.99491	15.999	O
S-sulphination	31.98983	31.999	O2
Farnesyl	204.18780	204.352	C15 H24
Geranylgeranyl	272.25040	272.469	C20 H32
Cysteic acid	47.98474	47.998	O3
Glutathionation	305.06816	305.308	H(15) C(10) N(3) O(6) S
Methyl methanethiosulfonate (MMTS)	45.98772	46.091	H(2) C S
N-ethylmaleimide (NEM)	125.04768	125.126	H(7) C(6) N O(2)
Carbamidomethyl (iodoacetamide)	57.02146	57.051	H(3) C(2) N O
Carboxymethyl (iodoacetic acid derivative)	58.00548	58.036	H(2) C(2) O(2)
S-pyridylethyl (4-vinylpyridine)	105.05785	105.137	H(7) C(7) N
Acrylamide (propionamide)	71.03711	71.078	C3 N O H5
HPDP-biotin	428.61240	428.613	H(32) C(19) N(4) O(3) S(2)

now considered a major PTM, playing key roles in a variety of important cellular functions.

Aberrant levels of protein S-nitrosylation have been associated with pathological conditions in humans such as strokes, Parkinson's disease, Alzheimer's disease, and amyotrophic lateral sclerosis [5]. In Arabidopsis, nitrosative stress caused by the inactivation of the denitrosylase enzyme, S-nitrosoglutathione reductase (GSNOR), leads to higher levels of protein S-nitrosylation and, as a consequence, higher susceptibility to pathogens [6].

One reason for the higher susceptibility of AtGSNOR mutants has been shown to be the impairment in salicylic acid signaling and synthesis [6–8]. Also, pathogen-triggered reactive oxygen intermediates (ROI) production may be regulated by S-nitrosylation. SNO formation within NADPH oxidase (AtRBOHD), a source of

ROI production, at Cys 890 appears to prevent the binding of the cofactor flavin adenine dinucleotide (FAD) to the active site of this enzyme, thus inhibiting its activity [9].

Another important target of *S*-nitrosylation in plant signaling is CDC48, which has been uncovered as a component in cryptogein-triggered NO signaling. In this context, Cys526 might function as a redox switch in the regulation of this protein [10]. Moving forward, it will be interesting to uncover the role of CDC48 in disease resistance and how *S*-nitrosylation of Cys526 might modulate this activity.

Salicylic acid (SA) is a key small molecule immune activator in plants [11]. SA has been shown to bind and modulate the activity of a number of proteins integral to plant immunity [12, 13]. SA-binding protein 3 (SABP3) exhibits a high affinity for SA and shows carbonic anhydrase (CA) activity [12]. CA activity is thought to be required for lipid biosynthesis [14] and lipids are thought to be important signals in plant immunity and have been linked to SA function [15]. Interestingly, SABP3 has been shown to be *S*-nitrosylated in vivo during the later stages of the plant defense response [7]. *S*-nitrosylation of SABP3 has been shown to be *S*-nitrosylated at Cys280 and this modification was directly proportional to AtGSNOR1 activity, which controls the cellular levels of SNOs. Furthermore, SABP3 *S*-nitrosylation at Cys280 diminished both SA binding and CA activity. Significantly, AtSABP3 was also found to be required for wild-type levels of disease resistance [7]. Collectively, these findings suggested that inhibition of AtSABP3 CA function by *S*-nitrosylation might contribute to a negative feedback loop that modulates plant immunity.

The emerging data also suggests that SNO function may have important roles outside of plant immunity [16]. Thus, *S*-nitrosylation also appears to regulate the phytohormone signaling pathways involved in plant development and adaptation to environmental stresses, such as cytokinin (CK) and abscisic acid (ABA), respectively. In this context, *S*-nitrosylation of the protein kinase open stomata 1 (OST1), a key component of ABA signaling pathway, at Cys137, has recently been shown to negatively regulate ABA function, impairing ABA-induced stomatal closure [17]. CK signaling has also been shown to be suppressed through *S*-nitrosylation of the histidine phosphotransfer protein 1 (AHP1) at Cys115 [18]. In this regard, identification and quantification of a given protein *S*-nitrosylation site is an essential step toward understanding the molecular mechanisms underpinning the given modification. However, the highly labile and redox sensitive SNO-cysteine bond makes this identification difficult.

With the advent of the biotin-switch technique [19], which selectively replaces NO moiety of SNO-cysteine residues by biotin-HPDP (*N*-[6-(biotinamido)hexyl]-3'-(2'-pyridyldithio)-propionamide), combined with mass spectrometry technologies, a large number of targets for *S*-nitrosylation have been characterized.

Although these techniques have greatly contributed to moving the field forward, optimizations are frequently, if not always, necessary for each protein of interest. In addition, several issues with the use of a biotin-HPDP based switch are frequently encountered such as the large size of the chemical moiety attached to the peptide containing cysteine which can interfere with the LC–MS separation, ionization, and fragmentation of the peptide.

Here, we describe an optimized method successfully used in our laboratory for the identification of *S*-nitrosothiols in recombinant proteins of importance in the immune responses of plants and animals. This method, described in details in Subheading 3.1, uses reagents that form thioether instead of disulphide bonds in both blocking and labeling steps, which, unlike the traditionally employed reagents, MMTS for blocking, and Biotin-HPDP for labeling, irreversibly modify thiols. The advantage of irreversibly modifying thiols is that samples can be completely denatured and reduced by strong reducing agents before trypsin treatment, thus, enhancing tryptic digestion and downstream LC–MS/MS sensitivity, described in detail in Subheading 3.2. A similar strategy of sequentially blocking cysteine has been reported by Camerini [20], here we broaden the applications of this method.

2   Materials

2.1   Reagents

1. Double-distilled water (ddH$_2$O).
2. 100 mM in 500 mM HCl: l-cysteine.
3. 100 mM in ddH$_2$O: Sodium nitrite (NaNO$_2$).
4. 500 mM in water: HEPES pH 7.8.
5. 500 mM in water: EDTA pH 7.8.
6. 200 mM in methane: Neocuproine.
7. 2.5 M in dimethyl sulfoxide (DMSO): *N*-ethylmaleimide (NEM).
8. 250 mM HEPES, 1 mM EDTA, 0.1 mM Neocuproine:HEN buffer pH 7.8.
9. 10 % SDS (w/v) in ddH$_2$O.
10. 25 % SDS (w/v) in ddH$_2$O.
11. 250 mM HEPES, 1 mM EDTA, 0.1 mM Neocuproine, 5 % (w/v) SDS, 50 mM NEM: Blocking buffer.
12. 70 % Acetone in ddH$_2$O.
13. 250 mM HEPES, 1 mM EDTA, 0.1 mM Neocuproine,1 % (w/v) SDS:HENS$_{1\%}$.

14. 500 mM in ddH$_2$O: Sodium ascorbate.

15. 100 mM in ddH$_2$O: Iodoacetamide.

16. 25 mM HEPES, 1 mM EDTA, 0.1 mM Neocuproine, 1 % (w/v) SDS :H$_{25}$ENS$_{1\%}$.

17. 1 M in ddH$_2$O: Dithiothreitol (DTT).

18. 200 mM Tris–Cl pH 6.8, 8 % (w/v) SDS, 0.4 % (w/v) bromophenol blue, 40 % glycerol :4× loading buffer.

19. 0.2 % coommassie brilliant blue R250 in 40 % methanol, 10 % glacial acetic acid: Staining solution.

20. 20 % methanol, 10 % acetic acid: Destaining solution.

21. Acetonitrile (ACN).

22. Ammonium bicarbonate (ABC).

23. 200 mM ABC, 50 %: ACN ABC200/ACN50.

24. 20 mM DTT, 200 mM ABC, 50 % ACN: DTT20/ABC200/ACN50.

25. 40 mM ABC, 9 % ACN: ABC40/ACN9.

26. 97.5 % H$_2$O, 2.5 % acetonitrile, 0.1 % formic acid: Mobile phase A.

27. 90 % acetonitrile, 10 % H$_2$O, 0.025 % trifluoroacetic acid, 0.1 % formic acid: Mobile phase B.

28. Capillary Picotip columns (10 cm × 360 μm outside diameter, OD, ×75 μm internal diameter, ID) with a 15 μm tip opening from New Objective.

29. Fused silica of different ID sizes (20 μm for splitflow, 50 μm as transfer line, and 200 μm for precolumn), the OD size used was fixed at a single size of 360 μm which is easier to secure using standard ferrules and fitting (VALCO and Upchurch scientific).

30. Access to a HPLC-tandem high mass accuracy mass spectrometer (Agilent 1200-Orbitrap LTQ XL for example).

31. File convertor tool like MSconvert from Proteomwizard.

32. Access to a search engine to identify peptide such as Mascot (Matrix Science), Sequest (ThermoFisher), or free application such as GPM (www.thegpm.org/), Maxquant (http://www.maxquant.org/links.htm).

33. Access to a tool to perform MS quantitation such as Progenesis (Nonlinear Dynamics), IDEAL-Q (http://ms.iis.sinica.edu.tw/IDEAL-Q/), or Maxquant to name a fews.

2.2 Specific Equipment

1. Standard peptide MRFA (Sigma, UK).

2. Progenesis (Nonlinear Dynamics, UK).

3 Methods

3.1 The Sequential Cysteine Blocking Technique

1. Prepare a 50 mM stock solution of nitric oxide donor CysNO by mixing 1:1 volumes of 100 mM l-cysteine and 100 mM sodium nitrite. Keep CysNO on ice and well protected from light exposure.

2. Exchange the buffer of your purified protein by passing it through Zeba spin desalting columns preequilibrated with HEN buffer pH 7.8 (follow Zeba's instructions to equilibrate) and, also using HEN buffer pH 7.8, dilute protein to a final concentration of 20 ng/μL (see **Note 1**).

3. Prepare three 1.5 mL microcentrifuge tubes by wrapping them with foil and marking as follow: negative control, sample, and positive control. To each tube, add 90 μL of protein. To the negative control tube, add 10 μL of HEN buffer pH 7.8. To the sample tube, add 9.8 μL of HEN buffer pH 7.8 and 0.2 μL of 50 mM CysNO (200 μM final concentration). To the positive control tube, add 8.8 μL of HEN buffer pH 7.8, 0.2 μL of 50 mM CysNO (200 μM final concentration) and 1 μL of 10 % SDS (0.1 % final concentration). Quickly vortex the tubes and keep them protected from light at room temperature for 20 min (see **Note 2**). In the meantime, equilibrate new Zeba spin desalting columns with HEN buffer pH 7.8 and prepare the blocking buffer.

4. Add two volumes of Blocking buffer (200 μL) to the bottom of three new 1.5 mL microcentrifuge tubes, mark them appropriately (negative, sample, and positive) and after that place the preequilibrated Zeba columns on them (see **Note 3**).

5. Add samples to the Zeba columns and spin them down at $1500 \times g$ for 2 min at room temperature. After that, discard columns, immediately wrap tubes in foil, vortex, and transfer them to 50 °C heating block or water bath for 20 min. Vortex tubes every 5 min (see **Note 4**).

6. Precipitate samples by adding 2 volumes (400 μL) of cold (−20 °C) 100 % acetone and place tubes at −20 °C for 20 min.

7. Centrifuge tubes at $>15,000 \times g$ for 5 min at 4 °C, discard supernatant, and wash samples 3× with 500 μL of cold 70 % acetone. Protecting from light exposure, let the acetone evaporate for 5 min before proceeding to next step (see **Note 5**).

8. Add 85 μL of HENS$_{1\%}$ buffer to each tube and keep them at room temperature for 5–10 min for the pellet to rehydrate. Make sure samples are always protected from light exposure at this step Vortex samples extensively to resuspend the proteins in the solution.

9. Add 5 μL of freshly prepared 500 mM Sodium Ascorbate (25 mM final concentration) and 10 μL of freshly prepared 100 mM Iodoacetamide (1 mM final concentration), quickly vortex the tubes, and incubate on a rocker for 1 h. Protect samples from light exposure.

10. Precipitate and wash samples as described in **steps 6** and **7**. After this point, light exposure is not an issue.

11. Resuspend samples in 19.5 μL of $H_{25}ENS_{1\%}$, add 3 μL of 1 M DTT, and 7.5 μL of 4× loading buffer.

12. Heat samples at 90 °C for 10 min and run them in an SDS PAGE gel with appropriate percentage of acrylamide (*see* **Note 6**).

13. Remove SDS PAGE gel from plates and rinse it with ddH$_2$O. Stain the gel by replacing the water with a staining solution and incubating it on a rocking plate for 30 min. To destain the gel, change the solution to a destaining solution, incubate on a rocking plate for 60 min. Replace the destaining solution by a fresh one and incubate overnight.

14. Rinse the destained gel 3× with ddH$_2$O and, using a clean blade, excise the bands corresponding to your samples and divide each of them in four small pieces. Transfer each piece to a 1.5 mL microcentrifuge tube and proceed to the next step (*see* **Note 7**).

15. Add 300 μL of ABC200/ACN50 solution and incubate samples at room temperature for 30 min. Repeat this step 3×.

16. Replace ABC200/ACN50 solution by 300 μL of DTT20/ABC200/ACN50 solution and incubate samples at room temperature for 1 h.

17. Wash gel pieces 3× with 500 μL of ABC20/ACN50 solution and remove solution from the tubes.

18. Cover gel pieces with 100 % ACN and incubate at room temperature for 5 min (*see* **Note 8**).

19. Remove ACN from the tubes and allow gel pieces to dry. In the meantime, prepare trypsin solution to a final concentration of 20 μg/mL by adding 100 μL of 1 mM HCl to a new 20 μg vial of trypsin, mix it well, and add 900 μL of ABC40/ACN9. Keep trypsin solution on ice.

20. Add 30 μL of trypsin solution to each gel piece and keep them at 4 °C until the gel pieces swell (*see* **Note 9**).

21. Incubate samples overnight at 37 °C. Next morning, using a water bath sonicator, sonicate sample for 5 min to maximize peptides release from the gel pieces to the solution.

22. Combine the digestions that correspond to the same samples in new 1.5 mL microcentrifuge tubes and discard the gel

pieces. The final volume per sample (negative control, sample, and positive control) should be approximately 120 μL (4×30 μL).

23. Concentrate samples by centrifuging the 1.5 mL microcentrifuge tubes in a low pressure centrifuge (speedVac concentrator) until the volume of the samples reduce to dryness. Proceed to mass spectrometry analysis or keep samples at –80 °C.

3.2 LC–MS/MS and Data Analysis

1. Resuspend samples in 10 μL of mobile phase A.

2. Using an Agilent micro-WPS auto sampler at 5 μL/min, inject 8 μL of sample into the HPLC–MS/MS apparatus (*see* **Note 10**).

3. After sample loading, reduce the flow rate across the column to approximately 100–200 nL/min using a vented column arrangement.

4. Analyze samples on a 60 min gradient for data dependant analysis. Use the follow solvent gradient program: 0 % buffer B (0–6 min), 0–5 % buffer B (6–8 min), 5–15 % (8–18 min) buffer B, 15–35 % buffer B (18–40 min), 35–100 % buffer B (40–48 min), followed by 100 % buffer B for 9 min and back to 0 % buffer B for 3 min. A column conditioning of 20 min with mobile phase A between samples is used. Optimize the MS instrument transmission using mostly the +1 charge state of the standard peptide MRFA in static nanospray mode. Set the ESI voltage at 1.7 kV and the interface temperature at 200 °C for all measurements.

5. In data dependant mode, the MS acquisition settings are detailed in **Note 11**.

6. For rapid identification, convert the RAW data to generate a Mascot compatible file by using the EXTRACT_MSN tool (Bioworks). A Mascot Generic File (MGF) will be generated and the data are then ready for Mascot search (Matrix science). Another strategy to generate an MGF file consists of using MSConvert, a functionality of ProteoWizard [8].

7. Search MS/MS data using MASCOT (Matrix Science Ltd, UK) against an appropriate database using a maximum missed-cut value of 2. Use variable methionine oxidation, cysteine carbamidomethylation and *N*-ethylmaleimide in all searches. For specific mass modification *see* Table 1.

8. Set the precursor mass tolerance to 7 ppm, the MS/MS tolerance to 0.4 amu and the significance threshold (p) below 0.05 (MudPIT scoring).

9. A peptide Mascot score of 20 is used in the final analysis to which we determine a global false discovery rate using a decoy database search.

10. For a simple identification based peptide, use the outcome of a standard search as describe in **step** 7, look for cysteine-containing peptides presenting a mass difference of 57.051 (Carbamidomethyl) for samples treated with CysNO and a mass difference of 125.126 (N-ethylmaleimide) for samples not treated with CysNO. A typical experimental design should follow the pattern: masses from untreated (negative control), CysNO (sample of interest), and CysNO plus SDS treated (positive control) sample.

11. A more in-depth analysis can be performed using peak intensity quantitation analysis as follows: Perform a Label-free quantification using Progenesis (Nonlinear Dynamics, UK). Reduce the total number of Features (i.e., intensity signal at a given retention time and *m/z*) to MS/MS peaks with charge of 2, 3, or 4+ and only keep the five most intense MS/MS spectra per "Feature."

12. Extract the subset of multi-charged ions (2+, 3+, 4+) from each LC–MS run and sum the ion intensities for normalization.

13. For calculating the *p*-value, it is preferable to work on transformed data. Thus, transform the measured peptide abundances using an ArcSinH function (as the method of detection can generate a significant amount of near zero measurements).

14. Calculate the within group means to determine the fold change and the transformed data to calculate the *p*-values using one way ANOVA.

15. Use ArcSinH transformation for the calculation of the *p*-value. Differentially expressed peptides are considered meaningful under the following conditions: only peptides detected with ratio of at least 1.5 (i.e., 1.5-fold upregulated or 0.667 downregulated) and $p < 0.05$ associated with the peptide change.

16. For the quantitative analysis from Progenesis, ensure the protein intensity (sum of intensity of all peptide composing this protein) in all sample conditions is the same, if not, although not ideal, normalize protein, and consequently peptide intensity.

4 Notes

1. This step is crucial if the original buffer (protein elution buffer) contained any reducing agents such as DTT or β-mercaptoethanol. It is very important to make sure that no reducing agents are present in the buffer. Small amounts of DTT or β-mercaptoethanol would reduce S-nitrosothiols and

interfere with final result. We use desalting columns because it is quicker and more convenient but alternatives to buffer exchange like dialyses may also be used.

2. S-nitrosothiols are also decomposed by ultraviolet light (UV). To avoid light interferences, CysNO should be added to the samples last and tubes should quickly be transferred to a place protected from light (like a drawer or a box). We also always switch lights off and close window blinds to reduce as much as possible the amount of light in the room.

3. The high SDS concentration and high temperature in the blocking step will expose free thiols that did not react with NO. Therefore, it is very important to remove all the CysNO before the samples are in contact with the blocking buffer so false-positives are avoided. It is also important to work quickly and reduce to the minimum the handling time of the samples to avoid light exposure, so no S-nitrosothiols are decomposed by UV. Thus, having the blocking buffer prepared and desalting columns in place make it quicker and easier.

4. Temperature, time of blocking, and concentration of blocking agent may vary depending on the given protein. If free thiols of a protein are not completely blocked at this step, they may generate false-positive signals. Therefore, conditions should be optimized for a protein when signals are detected in the negative control sample.

5. Protein pellets will not be visible. Add the acetone to the side of the tube opposite to where the pellet should be located and wash the pellet by inverting the tubes 3×. Centrifugations are not needed after each wash.

6. The ideal percentage of acrylamide will depend on the size of your protein of interest. Leaving an empty well in between the wells loaded with samples will make it easier and safer to load (avoiding cross-samples contaminations) and also to excise the bands from the gel afterward.

7. Avoid keratin contamination by wearing gloves all the time and avoid cross sample contamination by using one clean blade per sample. Gel pieces can be kept at −20 °C for months.

8. Gel pieces should turn white. If gel pieces do not turn white within 5 min wait a few minutes longer until they turn completely white.

9. Gel pieces should turn transparent again after a few minutes. It is important to keep the tubes at 4 °C until the gel pieces swell to avoid/reduce auto digestion of trypsin.

10. For micro-HPLC–MS/MS analysis, we use a micropump Agilent 1200 binary HPLC system coupled to a hybrid LTQ-Orbitrap XL instrument on a micro column-vented precolumn.

Capillary Picotip columns (10 cm × 360 μm OD × 75 μm ID) with a 15 μm tip opening and fitted with a borosilicate frit. The precolumn is a 4 cm × 360 μm OD × 200 μm ID. Frit are made in house using Kasil frit from Chembio LTD (76 μL Kasil, 32 μL formamide), mix thoroughly, dip the capillary rapidly and cure at 100 °C overnight, remove excess frit to only keep a 3 mm thick frit. Both column and precolumn are packed in-house using a pressure vessel with a slurry of 50 mg reverse phase packing material in 1 mL of chloroform. Fused silica tubing was purchased from Composite Metal (UK). The reverse phase bulk material is 5 μm Pursuit C18 obtained from Varian (UK).

11. A single FT scan at a resolution of 60 k in profile mode (400–2000 amu) is first performed with the lock mass function (background ion polysiloxane ion signal at 445.120025 amu), followed by five data dependent MS/MS scans of the five most intense ions in LTQ centroid mode with a window of 2 amu. Selected MS/MS precursor ion data are excluded for 180 s and the exclusion list is set at 500 for a narrow isolation width of 0.05 amu. The maximum fill time and target value are 200 ms, 200 ms, 1×10^6 and 1×10^4 ions for the FT survey scan and IT MS^2, respectively.

References

1. Yu M, Lamattina L, Spoel S, Loake G (2014) Nitric oxide function in plant biology: a redox cue in deconvolution. New Phytol 202(4): 1142–1156

2. Tsang A, Lee Y, Ko H (2009) S-nitrosylation of XIAP compromises neuronal survival in Parkinson's disease. Proc Natl Acad Sci 106(12): 4900–4905

3. Spoel S, Tada Y, Loake G (2010) Post-translational protein modification as a tool for transcription reprogramming. New Phytol 186(2):333–339

4. Sha Y, Marshall HE (2012) S-nitrosylation in the regulation of gene transcription. Biochim Biophys Acta 1820(6):701–711

5. Nakamura T, Tu S, Akhtar M, Sunico C (2013) Aberrant protein s-nitrosylation in neurodegenerative diseases. Neuron 78(4):596–614

6. Feechan A, Kwon E, Yun B-W, Wang Y, Pallas JA, Loake GJ (2005) A central role for S-nitrosothiols in plant disease resistance. Proc Natl Acad Sci U S A 102(22):8054–8059

7. Wang Y-Q, Feechan A, Yun B-W, Shafiei R, Hofmann A, Taylor P, Xue P, Yang F-Q, Xie Z-S, Pallas JA, Chu C-C, Loake GJ (2009) S-nitrosylation of AtSABP3 antagonizes the expression of plant immunity. J Biol Chem 284(4):2131–2137

8. Tada Y, Spoel SH, Pajerowska-Mukhtar K, Mou Z, Song J, Wang C, Zuo J, Dong X (2008) Plant immunity requires conformational changes [corrected] of NPR1 via S-nitrosylation and thioredoxins. Science 321(5891):952–956

9. Yun B-W, Feechan A, Yin M, Saidi NBB, Le Bihan T, Yu M, Moore JW, Kang J-G, Kwon E, Spoel SH, Pallas JA, Loake GJ (2011) S-nitrosylation of NADPH oxidase regulates cell death in plant immunity. Nature 478(7368): 264–268

10. Astier J, Besson-Bard A, Lamotte O, Bertoldo J, Bourque S, Terenzi H, Wendehenne D (2012) Nitric oxide inhibits the ATPase activity of the chaperone-like AAA+ ATPase CDC48, a target for S-nitrosylation in cryptogein signalling in tobacco cells. Biochem J 447(2):249–260

11. Loake G, Grant M (2007) Salicylic acid in plant defence--the players and protagonists. Curr Opin Plant Biol 10(5):466–472

12. Slaymaker DH, Navarre DA, Clark D, del Pozo O, Martin GB, Klessig DF (2002) The

tobacco salicylic acid-binding protein 3 (SABP3) is the chloroplast carbonic anhydrase, which exhibits antioxidant activity and plays a role in the hypersensitive defense response. Proc Natl Acad Sci U S A 99(18):11640–11645

13. Kumar D, Klessig DF (2003) High-affinity salicylic acid-binding protein 2 is required for plant innate immunity and has salicylic acid-stimulated lipase activity. Proc Natl Acad Sci U S A 100(26):16101–16106

14. Hoang CV, Chapman KD (2002) Biochemical and molecular inhibition of plastidial carbonic anhydrase reduces the incorporation of acetate into lipids in cotton embryos and tobacco cell suspensions and leaves. Plant Physiol 128(4):1417–1427

15. Kachroo P, Shanklin J, Shah J, Whittle EJ, Klessig DF (2001) A fatty acid desaturase modulates the activation of defense signaling pathways in plants. Proc Natl Acad Sci 98(16):9448–9453

16. Kwon E, Feechan A, Yun B-W, Hwang B-H, Pallas JA, Kang J-G, Loake GJ (2012) AtGSNOR1 function is required for multiple developmental programs in Arabidopsis. Planta 236(3):887–900

17. Wang P, Du Y, Hou Y-J, Zhao Y, Hsu C-C, Yuan F, Zhu X, Tao WA, Song C-P, Zhu J-K (2015) Nitric oxide negatively regulates abscisic acid signaling in guard cells by S-nitrosylation of OST1. Proc Natl Acad Sci U S A 112(2):613–618

18. Feng J, Wang C, Chen Q, Chen H, Ren B, Li X, Zuo J (2013) S-nitrosylation of phosphotransfer proteins represses cytokinin signaling. Nat Commun 4:1529

19. Jaffrey SR, Erdjument-Bromage H, Ferris CD, Tempst P, Snyder SH (2001) Protein S-nitrosylation: a physiological signal for neuronal nitric oxide. Nat Cell Biol 3(2):193–197

20. Camerini S, Polci M, Bachi A (2005) Proteomics approaches to study the redox state of cysteine-containing proteins. Ann Ist Super Sanita 41(4):451–457

Chapter 15

Detection of *S*-Nitrosoglutathione Reductase Activity in Plants

Lucie Kubienová, Tereza Tichá, Lenka Luhová, and Marek Petřivalský

Abstract

S-nitrosoglutathione reductase (GSNOR) is considered a key enzyme in the regulation of intracellular levels of *S*-nitrosoglutathione and protein *S*-nitrosylation. As a part of nitric oxide catabolism, GSNOR catalyzes the irreversible decomposition of GSNO to oxidized glutathione. GSNOR is involved in the regulation of plant growth and development, mediated by NO-dependent signaling mechanisms, and is known to play important roles in plant responses to various abiotic and biotic stress conditions. Here we present optimized protocols to determine GSNOR enzyme activities in plant samples by spectrophotometric measurements and by activity staining after the native polyacrylamide gel electrophoresis.

Key words *S*-nitrosylation, *S*-nitrosothiols, Nitric oxide, *S*-nitrosoglutathione reductase, Plant stress

1 Introduction

Nitric oxide (NO) is considered as an important signaling molecule in a plethora of biological processes in a wide range of organisms. NO is involved in the regulation of plant developmental, growth processes, and signaling cascades of plant responses to stress conditions [1]. *S*-nitrosothiols are formed by covalent modification of cysteine thiol with nitroso (NO–) group both in proteins and low-molecular-weight thiols. *S*-nitrosylation is currently considered as one of the most important posttranslational protein modifications and increasing number of *S*-nitrosylation targets has been identified among proteins and enzymes involved in crucial cellular processes [2]. *S*-nitrosoglutathione (GSNO) represents the major low-molecular-weight *S*-nitrosothiol, which is capable under specific conditions to release free NO or to participate in NO transport and in transnitrosation reactions, when nitroso group is transferred to thiol group of another molecule [3]. Similarly to animal cells, GSNO is supposedly the most abundant low-molecular-weight *S*-nitrosothiol in plant cells, with high significance to the signaling pathways and metabolism of NO in plants

Kapuganti Jagadis Gupta (ed.), *Plant Nitric Oxide: Methods and Protocols*, Methods in Molecular Biology, vol. 1424,
DOI 10.1007/978-1-4939-3600-7_15, © Springer Science+Business Media New York 2016

[4]. *S*-nitrosoglutathione reductase (GSNOR) plays a crucial role in GSNO catabolism in the regulation of protein *S*-nitrosothiol levels and in the metabolism of NO and reactive nitrogen species, as reviewed elsewhere [5].

GSNOR activity is associated with another previously known enzyme, formerly referred to as glutathione-dependent formaldehyde dehydrogenase (FALDH; EC 1.2.1.1). *S*-(hydroxymethyl) glutathione (HMGSH) as a spontaneous adduct of glutathione and formaldehyde was later identified as the proper substrate; therefore the enzyme was reclassified as *S*-(hydroxymethyl)glutathione dehydrogenase (EC 1.1.1.284). It catalyzes the oxidation of HMGSH to *S*-formylglutathione as a part of formaldehyde detoxification [6]. However, the NADH-dependent reduction of GSNO, leading to the formation of oxidized glutathione (GSSG) and ammonium, has been uncovered as a more physiologically relevant reaction [7, 8]. For this reason, the denomination of this enzyme as GSNOR is currently highly extended within scientific literature, although this designation has not been accepted yet by IUBMB nomenclature commission. According to the enzyme classification, GSNOR belongs to a family of Zn-dependent class III alcohol dehydrogenases (ADH3; EC 1.1.1.1), which are distinct from members of class I ADH family by their high affinity towards long-chain alcohols like cinnamyl alcohol, farnesol, and geraniol [9]. *S*-nitrosoglutathione is considered as the physiologically most relevant substrate, as the K_m value of GSNOR for HMGSH is about 100 times higher as compared to GSNO [10].

GSNOR has important role in the regulation of *S*-nitrosoglutathione levels in plants during their development and under stress conditions [11]. Plant GSNORs are ubiquitously expressed cysteine-rich cytosolic proteins involved in the regulation of NO homeostasis, shoot morphology, and pathogen defence responses [12]. Activity of *S*-nitrosoglutathione reductase is assessed by measuring the amount of NADH consumed during the enzyme reaction monitoring the decrease of absorbance at 340 nm. Inversely in the dehydrogenase mode, the oxidation of suitable substrate such as HMGSH is followed by the increase of absorbance at 340 nm. Electrophoretic separation under non-denaturing conditions (native polyacrylamide gel electrophoresis, native PAGE) is also often used to estimate enzyme activities and to analyze isoenzyme spectra in plant extracts. Fluorescence method for the detection of GSNOR reductase activity in the native PAGE gels is based on the detection of decreased fluorescence signal (excitation 340 nm, emission 460 nm) due to the consumption of NADH cofactor. Gels are sequentially incubated with filter papers soaked in phosphate buffers containing NADH and then GSNO. After short incubation, GSNOR enzymes are detected as dark bands on exposed gels [13, 14]. Alternatively, staining for GSNOR dehydrogenase activity can be performed using electron acceptors such as nitroblue tetrazolium (NBT). Reduction of tetrazolium compounds leads to the formation of colored bands of insoluble formazan produced at the site of GSNOR presence in the gels. HMGSH, prepared by mixing of corresponding solution of reduced glutathione and formaldehyde, or long-chain alcohols like octanol, is used as GSNOR substrate [15]. The advantages of using dehydrogenase GSNOR staining in native PAGE gels include higher stability of used compounds and easier detection of visible bands without the need of more expensive equipment necessary for the detection of NADH fluorescence changes.

2 Materials

Prepare all solution using ultrapure and chemicals of analytical grade. Store all reagents at room temperature, unless indicated otherwise. Follow waste regulation to appropriately dispose waste material.

2.1 Reagents

1. 1 M HCl.
2. 614 mg Reduced GSH.
3. 138 mg Sodium nitrate.
4. 10 ml Acetone.
5. 10 ml Diethyl ether.
6. 1 mM Benzenesulfonyl fluoride hydrochloride.
7. 50 mM Tris–HCl pH 7.5.
8. 10 mg/ml BSA solution.

2.2 Determination of Protein Content by Bradford Method

1. Coomassie Brilliant Blue (solution: weight 50 mg of Coomassie Brilliant Blue G250, dissolve it in 25 mL of methanol, add 50 mL of 85 % H_3PO_4, and make up to 100 mL with water) (*see* **Note 1**).
2. Protein standard: 10 mg/mL Bovine serum albumin (BSA) in water. Aliquots can be stored at –20 °C.

2.3 Determination of GSNOR Reductase Activity

1. Reaction buffer: 20 mM Tris–HCl, pH 8.0. Weight 2.423 g Tris, dissolve in 950 mL of water, mix, adjust pH with HCl, and make up to 1 L with water. Store at 4 °C.
2. Reduced nicotinamide adenine dinucleotide (NADH): 2 mM NADH. Weight 1.42 mg of NADH disodium salt hydrate (Sigma-Aldrich, N8129) and dissolve in 1 mL of water. Always prepare fresh.
3. *S*-nitrosoglutathione: 4 mM GSNO. Weight 1.34 mg GSNO and dissolve in 1 mL of water. Always prepare fresh and keep prepared solutions on ice protected from the light (*see* **Note 2**).

2.4 Determination of GSNOR Dehydrogenase Activity

1. Reaction buffer: 20 mM Tris–HCl, pH 8.0. Weigh 2.423 g of Tris, dissolve in 950 mL of water, mix, adjust pH with concentrated HCl, and make up to 1 L with water. Store at 4 °C.

2. Nicotinamide adenine dinucleotide (NAD$^+$): 60 mM NAD$^+$. Weigh 40 mg NAD$^+$ hydrate (Sigma-Aldrich, N1636) and dissolve it in 1 mL of water. Always prepare fresh.

3. Glutathione: 40 mM. Weight 123 mg of GSH and dissolve it in 10 mL of water.

4. Formaldehyde: 31.2 mM. Add 31 μL of 30 % formaldehyde (about 10 M) to 10 mL of water (see **Note 3**).

2.5 Determination of In-Gel GSNOR Activity

0.1 M Sodium phosphate buffer, pH 7.4, containing 2 mM NADH or 4 mM GSNO: Weigh 7 mg NADH and dissolve in 5 mL 0.1 M sodium phosphate buffer, pH 7.4. Weigh 6.5 mg GSNO and dissolve in 5 mL 0.1 M sodium phosphate buffer, pH 7.4. Prior to the use, keep both solutions of NADH and GSNO at 4 °C protected from the light.

1. 0.1 M Sodium phosphate buffer (pH 7.4) containing 0.1 mM NAD$^+$, 0.1 mM nitroblue tetrazolium (NBT), 0.1 mM phenazine methosulfate (PMS), 1 mM reduced glutathione (GSH), and 1 mM formaldehyde (see **Note 4**). Weigh 6.6 mg NAD$^+$, 8.2 mg NBT, 3.1 mg PMS, and 30.7 mg GSH. Dissolve prepared compounds sequentially in given order in 100 mL 0.1 M sodium phosphate buffer, pH 7.4. Finally add 30 μL of formaldehyde. Store in the dark prior to use.

2.6 Native Polyacrylamide Gels

1. Resolving gel buffer: 1.5 M Tris–HCl, pH 8.8. Weigh 36.3 g Tris and dissolve in 150 mL of water. Adjust pH with HCl and make up to 200 mL with water. Store at 4 °C.

2. Stacking gel buffer: 0.5 M Tris–HCl, pH 6.8. Weigh 6.1 g Tris and dissolve in approx. 80 mL of water. Adjust pH with HCl and make up to 100 mL with water. Store at 4 °C.

3. Thirty percent acrylamide/bisacrylamide solution (30:0.8 acrylamide/bis): Weigh 30 g of acrylamide monomer and 0.8 g Bis (cross-linker) and transfer to a glass beaker. Add water to a volume of 60 mL and mix for about 30 min (see **Note 5**). Make up to 100 mL with water. Store at 4 °C protected from light.

4. Ammonium persulfate (APS): 10 % solution in water. Weigh 25 mg of APS and dissolve in 250 μL of water. Always prepare fresh (see **Note 6**).

5. N,N,N′,N′-tetramethylethylene diamine (TEMED). Store at 4 °C.

6. 60 % Glycerol (v/v) in water. Store at 4 °C.

7. 0.02 % Bromophenol blue in 20 % glycerol (v/v). Store at 4 °C

8. Native PAGE running buffer: 25 mM Tris–HCl, 0.192 M glycine, pH 8.3. Weigh 6.05 g Tris and 28.82 g glycine (electropho-

resis grade, e.g., Sigma-Aldrich G8898) and transfer to a glass beaker. Add water to a volume of 1 L, mix, and adjust pH with HCl. Make up to 2 L with water. Store at 4 °C (*see* **Note 7**).

3   Methods

All procedures shall be performed at room temperature unless otherwise specified.

3.1   Preparation of GSNOR Substrate S-Nitrosoglutathione (GSNO)

1. GSNO is synthetized by the nitrosylation of reduced glutathione (GSH) using acidified nitrite in HCl [16] (*see* **Note 8**).

2. HCl 1 M by adding dropwise 4.3 mL of concentrated HCl (36 %) to 40 mL of water and adjust final volume with water to 50 mL.

3. Dissolve reduced GSH 614 mg in 3 mL of 1 M HCl and place the solution on ice bath (*see* **Note 9**).

4. Slowly add 138 mg of sodium nitrite ($NaNO_2$) to the acidic glutathione solution while stirring vigorously on ice bath (*see* **Note 10**). Stir for at least 40 min on ice.

5. Add 5 mL of ice-cold acetone and stir for 10 min on ice to precipitate GSNO out of solution (*see* **Note 11**).

6. Filter out the solution from reaction mixture on Buchner funnel and wash the precipitate with successive portions of ice-cold water (two times 10 mL), ice-cold acetone (two times 10 mL), and ice-cold diethyl ether (two times 10 mL).

7. Dry out the final GSNO powder wrapped in aluminum foil in desiccator under vacuum overnight. Store at –20 °C (*see* **Note 12**).

3.2   Sample Extraction and Desalting

Extraction buffer: 50 mM Tris–HCl pH 7.5, 0.2 % Triton X-100, 2 mM DTT, 1 mM (4-2-aminoethyl)benzenesulfonyl fluoride hydrochloride (AEBSF) (*see* **Note 13**). Weigh 0.606 g Tris and 0.2 g Triton X-100, dissolve in 75 mL of water, mix, and adjust pH with concentrated HCl. Make up to 100 mL with water and store at 4 °C. DTT and AEBSF are added freshly before the use. To prepare 20 mL of the extraction buffer, dissolve 6.2 mg DTT and 4.8 mg AEBSF in approx. 15 mL of the TRIS buffer, and then make up to 20 mL with water.

1. Equilibration and elution buffer: 20 mM Tris–HCl, pH 8.0. Weight 2.423 g Tris, dissolve in 950 mL of water, mix, adjust pH with HCl, and make up to 1 L with water. Store at 4 °C.

3.3   Preparation and Desalting of Plant Extracts

1. Homogenize plant samples with extraction buffer (1:2, w/v) using a mortar and pestle in liquid nitrogen.

2. Centrifuge crude extracts for 20 min at 16000×g at 4 °C. Transfer supernatants into clean microtubes and store at 4 °C.

3. Equilibrate desalting column (PD-10 or NAP-10 columns, GE Healthcare, USA) with at least 15 mL of equilibration buffer.

4. Add 1 mL of supernatant to the column and let the sample to enter the packed bed completely.

5. Add 2 mL of elution buffer and collect the elute. Store the elute at 4 °C (*see* **Note 14**).

3.4 Determination of Protein Content by Bradford Method

1. Protein standard solutions: dilution of 10 mg/mL BSA solution in the range of 0.05–1.4 mg/mL (*see* **Note 15**).

2. Bradford reagent: diluting the stock Coomassie Blue solution with water in ratio 1:4 (*see* **Note 16**).

3. To each microplate well, pipette 5 μL of BSA protein standard, extract sample, or blank. Elution buffer used in sample desalting should be used as a blank. Protein samples, standards, and blank are usually assayed at least in triplicates.

4. Add 195 μL of Bradford reagent, mix, and incubate for 10 min at room temperature (*see* **Note 17**).

5. Measure the absorbance at 595 nm and subtract the mean of blank absorbance values from the measured absorbance values of standards and samples.

6. Protein sample concentrations are determined using standard curves obtained by plotting absorbance values of BSA protein standards versus their concentrations.

3.5 Determination of GSNOR Reductase Activity

1. GSNOR reductase activity is assayed spectrophotometrically at 25 °C by monitoring the oxidation of NADH at 340 nm (*see* **Note 18**).

2. Fresh 2 mM NADH and 4 mM GSNO. Store at 4 °C in the dark.

3. For microplate assay, add 225 μL of Tris–HCl buffer, 15 μL of the desalted sample, and 30 μL of 2 mM NADH to 96-well microplates in triplicate.

4. The reaction is started by addition of 30 μL of 4 mM GSNO; as the blank use 30 μL of water (*see* **Note 19**).

5. Monitor the decrease of absorbance at 340 nm during 1–10-min interval (*see* **Note 20**).

6. The activity is expressed as nmol of NADH consumed per min and mg of protein, using extinction coefficient of NADH $\varepsilon = 6.22$ mM^{-1} cm^{-1} (*see* **Note 21**).

3.6 Determination of GSNOR Dehydrogenase Activity

1. GSNOR dehydrogenase activity is assayed spectrophotometrically at 25 °C by monitoring the formation of NADH at 340 nm with *S*-(hydromethyl)glutathione (HMGSH) as the enzyme substrate (*see* **Note 22**).

2. Fresh solutions of 60 mM NAD⁺, 40 mM reduced glutathione, and 31.2 mM formaldehyde. Store at 4 °C.

3. In a microtube, mix 1 mL of 40 mM GSH and 1 ml of 31.2 mM formaldehyde to get 2 ml of 20 mM HMGSH (*see* **Note 23**).

4. For microplate assay, add 205 µL of 20 mM Tris–HCl buffer, 15 µL of the desalted sample, and 10 µL of 60 mM NAD⁺ to 96-well microplate in triplicate.

5. The reaction is started by adding 15 µL of 20 mM HMGSH; as the blank use 15 µL of water.

6. Monitor the increase of absorbance at 340 nm during 1–10-min interval (*see* **Note 19**).

7. The activity is expressed as nmol NADH produced per minute and mg protein ($\varepsilon = 6.22$ mM⁻¹cm⁻¹) (*see* **Note 20**).

3.7 Native Polyacrylamide Gel Electrophoresis and GSNOR Activity Staining

1. The following instructions refer to the preparation of two 0.75 mm thick minigels on the Mini-PROTEAN® Tetra System (Bio-Rad, USA).

2. Before use, short plates and spacer plates are thoroughly cleaned with deionized water and finally with 95 % ethanol (*see* **Note 24**).

3. Assemble two pairs of short and 0.75 mm spacer plate and insert them into the gel-casting stand. To prevent gel leakage, check tightness of the casting plate assembly by pouring water between the plates. Remove the water and let the plates to completely dry.

4. Solution for resolving gels: Mix 2.5 mL of resolving buffer, 3.3 mL of 30 % acrylamide/bis mixture, and 4.1 mL water in a 50 mL conical flask (*see* **Note 25**).

5. Degas the solution for at least 10 min under vacuum and continuous stirring (*see* **Note 26**).

6. Add 10 µL of TEMED to the degassed solution under continuous stirring.

7. Add 100 µL of freshly prepared APS under continuous stirring and pour the solution immediately between assembled glass plates. The solution level should reach approximately 1 cm below the end of the comb (*see* **Note 27**).

8. Using Pasteur pipette overlay the resolving gel solution with n-butanol saturated with water and allow to stand for 30 min (*see* **Note 28**).

9. After the polymerization of resolving gel, remove butanol and rinse the gel surface with water. Remove the water and dry the area between the glass plates above the gel surface using a piece of filter paper.

10. Solutions for stacking gels: Mix 1.25 mL of resolving buffer, 0.65 mL of acrylamide mixture, and 3.05 mL water in a 50 mL conical flask (*see* **Note 25**).

11. Degas this solution for at least 10 min under vacuum and continuous stirring (*see* **Note 26**).

12. Add 10 μL of TEMED to the degassed solution under continuous stirring.

13. Add 50 μL of APS under continuous stirring and pour immediately the solution on the top of stacking gel between assembled glass plates. The level of the solution should reach the top of the short plates.

14. Immediately insert a 0.75 mm comb between the glass plates without introducing air bubbles and let the stacking gel to polymerize for 30 min (*see* **Note 27**).

15. Prepare protein samples by mixing desalted plant extracts with 60 % glycerol in ration 3:1 (v/v) (*see* **Note 29**).

16. When the polymerization of stacking gels is finished, take out the glass plates with cast polyacrylamide gels from the casting stand, but keeping the comb inside the glass plates.

17. Assemble the electrophoretic cell according to the manufacturer's instructions and insert the plates carrying the polyacrylamide gel facing the short plates inside the cell.

18. Fill assembled electrophoretic cell with prepared native PAGE electrode buffer to the level above the end of short plates, carefully remove the combs, and rinse the wells in stacking gel thoroughly with the aid of Pasteur pipette (*see* **Note 30**).

19. Check for air bubbles under the glass plates—if present, remove them using Pasteur pipette (*see* **Note 31**).

20. Using Hamilton syringe or gel-loader pipette tips, transfer 20 μL of prepared protein samples slowly into selected wells. Do not use outer wells for samples loading; instead they can be used to load small amount of bromophenol solution (up to 5 μL) to follow the progress of the electrophoretic run (*see* **Note 32**).

21. Place the lid on the electrophoretic cell, connect the cables into to a suitable power supply, and set the power supply to 100 V to start the electrophoretic separation (*see* **Note 33**).

22. After the bromophenol blue zone reached the limit of separation gel, increase the voltage to 150 V. Continue the separation until the bromophenol blue migrates out from the gel. Then turn off the power supply and disconnect the cables from the power supply.

23. Disassemble the cell, pull out the glass plates with the gel, and gently remove the gel by separating the glass plates. Cut and

discard the stacking gel and mark the upper part of the running gel by cutting one of its edges (*see* **Note 34**).

24. If required, separated proteins can be detected in the polyacrylamide gels by any established method of total protein staining like Coomassie Blue or silver staining.

3.8 Detection of the GSNOR Activity in Gels

1. Four 8×5 cm pieces of thick (~1 mm) filter paper (*see* **Note 35**).

2. After the end of electrophoretic separation, remove the stacking gels and transfer with caution the separating gels to a suitable plastic tray (*see* **Note 36**).

3.8.1 Fluorescence Detection of GSNO Reductase Activity

3. Rinse gels for 5 min in deionized water. Place washed gels on the surface of UV tray for fluorescence imaging (*see* **Note 37**).

4. Soak two filter papers in 0.1 M sodium phosphate buffer, pH 7.4, containing 2 mM NADH (*see* **Note 38**).

5. Place soaked filter paper on the surface of the gel. Remove the air bubbles between the gel and filter paper rolling over a glass bar.

6. Incubate the gels in the dark for 15 min and then remove the filter papers soaked with NADH.

7. Take two filter papers and soak them with 0.1 M sodium phosphate buffer, pH 7.4, containing 4 mM GSNO (*see* **Note 38**).

8. Place soaked filter paper on the surface of the gel. Remove the air bubbles between the gel and filter paper using a glass bar.

9. Incubate the gels in the dark for 15 min and then remove the filter papers.

10. Place immediately the tray inside the documentation system (*see* **Note 39**).

11. Set the parameters for the exposure using suitable UV excitation and filter settings and start the imaging (*see* **Note 40**).

12. Finally, the presence of GSNOR activity appears as dark bands on the gel, which corresponds to the disappearance of NADH fluorescence.

13. If required, a semiquantitative comparison of signal intensities of detected bands can be performed using available software for image analysis (proprietary software of documentation system or any of freely available software as Image Lab).

14. An example of the analysis of GSNOR reductase activity in native PAGE gel is shown in Fig. 1a.

3.8.2 Staining for GSNOR Dehydrogenase Activity

1. Take suitable plastic tray and place the gel inside.

2. Pour in 50 mL of freshly prepared staining solution the box and incubate for 45 min at room temperature in the dark (*see* **Note 41**).

3. After the incubation, remove the staining solution, wash the gel shortly twice with water, and transfer the gel on White tray (*see* **Note 37**).

4. Colored formazan produced by GSNOR dehydrogenase activity forms purple-blue bands (*see* **Note 42**).

5. Place the tray inside a suitable documentation system, set the parameters for the exposure by white light, and start imaging (*see* **Note 40**).

6. If required, a semiquantitative comparison of signal intensities of detected bands can be performed using available software for image analysis (proprietary software of documentation system or any of freely available software as Image Lab).

7. An example of the analysis of GSNOR dehydrogenase activity in native PAGE is shown in Fig. 1b.

4   Notes

1. Stock Coomassie Blue solution should have reddish-brown color. It can be stored for months in the dark at 4 °C.

2. Fresh GSNO solution should have a pink color. A stock solution can be prepared by dissolving GSNO in an organic solvent like DMSO purged with an inert gas. Aqueous GSNO solution should not be stored for more than 1 day.

3. Commercial concentrated formaldehyde solutions are stable only until opening the bottle. In contact with air, formaldehyde solutions oxidize to formic acid and eventually polymerize to paraformaldehyde. It is therefore advisable to use formaldehyde from recently opened ampules. Alternatively, "fresh" formaldehyde solutions can be obtained by heating of paraformaldehyde.

4. The specificity of GSNOR activity staining can be checked by a parallel staining of another gel in a staining solution without glutathione addition.

5. Caution: Acrylamide is a neurotoxin. Always wear gloves, safety glasses, and a surgical mask when working with acrylamide powder. To avoid exposing your lab mates to acrylamide, work in the fume hood.

6. The use of freshly prepared ammonium persulfate solution is crucial for the initiation of acrylamide monomer polymerization to form the desired gels.

7. Running buffer for native PAGE can be reused several times, if stored at 4 °C.

8. Due to high GSNO sensitivity to light, whenever possible all procedures with GSNO compound and solutions should be performed avoiding exposure to light.

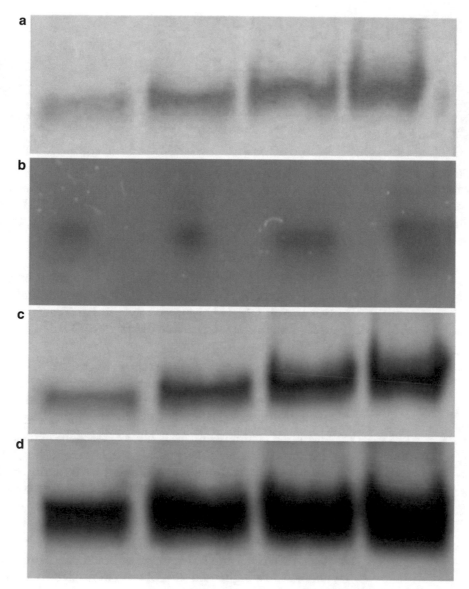

Fig. 1 Detection of enzyme activity of purified GSNOR protein. In each well 5, 10, 15, or 20 μg of purified recombinant GSNOR from tomato (*Solanum lycopersicum* cv. Amateur), prepared as described elsewhere [10], was loaded and proteins were separated by native PAGE. (**a**) Total protein staining by Biosafe reagent (BioRad), (**b**) reductase activity detected by fluorescence method, (**c**) dehydrogenase activity staining using NBT, (**d**) dehydrogenase activity staining using MTT.

9. Reduced GSH store at room temperature or at 4 °C undergoes slow oxidation by air oxygen. The use of older and inappropriately stored GSH results in decreased amounts of the reaction product.

10. The solution will turn red and intensive production of bubbles can be observed.

11. The best results in the precipitation and subsequent steps are obtained using acetone and ether previously kept for 1 h in the freezer at −20 °C (Fig. 2).

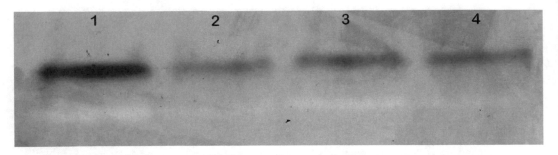

Fig. 2 Detection of GSNOR dehydrogenase activity in plant extracts. Samples were collected from 7-day-old pea seedlings grown under normal light regime or in the dark. Sample preparation, native PAGE electrophoresis, and staining for dehydrogenase activity using MTT were done as described in Methods. (*1*) Green seedlings—hypocotyl, (*2*) green seedlings—roots, (*3*) etiolated seedlings—hypocotyls, (*4*) etiolated seedlings—roots.

12. The quality of synthesized GSNO can be verified by spectroscopic comparison to commercial GSNO sample (Calbiochem, San Diego, CA, USA) measuring absorbance at 334 nm and using value of extinction coefficient $\varepsilon = 800$ mM-1 cm-1 for GSNO.

13. To inhibit proteases in the plant extract, AEBSF is recommended as a less toxic and more stable alternative to phenylmethylsulfonyl fluoride (PMSF). It can be purchased also under the commercial name Pefabloc© (Roche Diagnostic).

14. Desalting columns are available as disposable consumables, but our experience shows taking proper care during repeated use is feasible without compromising the column efficiency. After elution step, add at least 25 mL of water to clean the column before next desalting cycle. For long-term storage, add at least 15 mL of 20 % ethanol to avoid microbial contamination, and store at 4 °C.

15. This is the usual applicable range of Bradford method. Extract samples with higher amount of protein needed to be diluted.

16. Working Bradford reagent shall have bluish brown color.

17. A known drawback of the Bradford method is the long-term instability of color changes resulting from complexes formed with Coomassie Blue, namely in samples with high protein concentration. For this reason, it is highly recommended to read absorbance values less than 20 min after the addition of Bradford reagent to protein samples.

18. EDTA or similar chelating agents should not be used during neither sample extraction nor activity determination, as this can result in the removal of zinc atoms from GSNOR molecule and loss of activity. Chelating agents like EDTA completely inactivate the enzyme in the millimolar range. Reducing agents like β-mercaptoethanol, dithiothreitol, or

ascorbic acid inhibit both the reductase and dehydrogenase reactions of GSNOR, and at 1 mM concentration they reduce the enzyme activity by about 30 % [10].

19. Each sample must have own blank.

20. This time interval is usually optimal for measuring GSNOR in plant extracts. Analysis of samples with high GSNOR activity (i.e., purified recombinant proteins) may require shortened measuring time to obtain linear progress of the initial rate of enzyme reaction.

21. For the calculation of concentration changes using Lambert-Beer equation, the optical path length of the measured sample needs to be determined. If this option is not available on the used microplate reader, the height of the solution in the microplate wells can be calculated using the known volume and the well surface, which should be available from the microplate manufacturer.

22. Other GSNOR substrates are also used for the determination of dehydrogenase activity, namely long-chain alcohols as octanol or geraniol. In this case, the reaction buffer needs to have pH adjusted to 10.

23. S-(hydroxymethyl)glutathione (HMGSH) is a nonenzymatically formed adduct of glutathione (GSH) and formaldehyde [17]. Glutathione and formaldehyde concentrations required for the preparation of 1 mM HMGSH were calculated using the equation and K_{eq} value as described previously [10, 18].

24. Preparation of two identical gels is suitable for the detection of GSNOR in one gel and parallel detection of total proteins by Coomassie Blue, silver staining, or other available method in the other gel.

25. Buffers and acrylamide solution are usually stored at 4 °C. It is hence important to let the solution to room temperature before the preparation of gel solution, as the low temperature is known to have negative effect on the acrylamide polymerization and the quality of resulting gels.

26. Acrylamide polymerization involves free radical reactions, which are inhibited by oxygen or other compounds which has free radical scavenging properties. Solution degassing is therefore substantial to remove oxygen dissolved in the used solutions and buffers or adhered to the surfaces.

27. Acrylamide polymerization in the gel is initiated as soon as ammonium persulfate is added to the solution, so all subsequent actions must be performed promptly.

28. Butanol layer on the top of the gel limits the access of the air, thus accelerating the gel polymerization.

29. Glycerol is mixed with desalted plant extracts to increase the density of sample solutions to enable their easy transfer to wells in the stacking gels.

30. The role of electrode buffer is to provide electric connection between the upper and lower ends of the gel to generate electrophoretic movement of proteins in analyzed sample.

31. Air bubbles below the plates may result in irreproducible results due to irregular changes of the electric current passing through the gels.

32. The used amount of bromophenol solution should be as low as possible; otherwise bromophenol can eventually diffuse to the lanes of adjacent samples and affect the activity of present proteins.

33. Using electrophoresis power supplies, you should follow general rule of using electric appliances. Never connect the lid and switch on the power supply before the lid is securely placed on the electrophoretic cell.

34. Cutting one of the gel edges will help you to recognize the correct gel orientation in following steps of protein blotting.

35. For this purpose we successfully use paper available for the germination of plant seeds, which is an option far less expensive than original blotting papers provided by BioRad or other suppliers.

36. For all manipulation with PAGE gels and blotting membranes, always wear quality gloves.

37. Alternatively, place gels on a piece of glass, such as old glass plates for bigger gels.

38. Recommended volume is at least 2 mL for one piece of filter paper.

39. When using a glass plate as a gel support, place it over the UV illuminator of the documentation system.

40. Exposure settings need to be optimized for each experiment and used documentation system.

41. All solutions should be prepared fresh and protected from the light. Moreover, the staining solution should be thoroughly mixed after the addition of each compound and we recommend to add the compounds in the given order. If required, the method sensitivity can be increased by incubation of the gels with staining solution at 37 °C in the dark.

42. The resulting color will depend on the choice of tetrazolium salt used for the staining. Generally, we observed comparable results using NBT or MTT to detect GSNOR dehydrogenase activity.

For some samples MTT gave more intense staining; this was however accompanied by more intense background staining of the whole gel compared to NBT.

Acknowledgments

This project was supported by the grants Kontakt LH11013 and GAČR P501/12/0590. We thank Prof. Juan B. Barroso (Universidad of Jaén, Spain) and Dr. Javier F. Corpas (CSIC Granada, Spain), for valuable collaboration and sharing practical knowledge in the initial steps of our research in plant NO and GSNOR field.

References

1. Yu M, Lamattina L, Spoel SH, Loake GJ (2014) Nitric oxide function in plant biology: a redox cue in deconvolution. New Phytol 202:1142–1156
2. Seth D, Stamler JS (2011) The SNO-proteome: causation and classifications. Curr Opin Chem Biol 15:129–136
3. Martínez-Ruiz A, Lamas S (2004) S-nitrosylation: a potential new paradigm in signal transduction. Cardiovasc Res 62:43–52
4. Corpas FJ, Alché JD, Barroso JB (2013) Current overview of S-nitrosoglutathione (GSNO) in higher plants. Front Plant Sci 4:126
5. Benhar M, Forrester MT, Stamler JS (2009) Protein denitrosylation: enzymatic mechanisms and cellular functions. Nat Rev Mol Cell Biol 10:721–732
6. Koivusalo M, Baumann M, Uotila L (1989) Evidence for the identity of glutathione-dependent formaldehyde dehydrogenase and class III alcohol dehydrogenase. FEBS Lett 257:105–109
7. Jensen D, Belka G, Du Bois G (1998) S-Nitrosoglutathione is a substrate for rat alcohol dehydrogenase class III isoenzyme. Biochem J 331:659–668
8. Liu L, Hausladen A, Zeng M et al (2001) A metabolic enzyme for S-nitrosothiol conserved from bacteria to humans. Nature 410:490–494
9. Staab CA, Hellgren M, Höög JO (2008) Medium- and short-chain dehydrogenase/reductase gene and protein families. Cell Mol Life Sci 65:3950–3960
10. Kubienová L, Kopečný D, Tylichová M et al (2013) Structural and functional characterization of a plant S-nitrosoglutathione reductase from Solanum lycopersicum. Biochimie 95:889–902
11. Leterrier M, Chaki M, Airaki M et al (2011) Function of S-nitrosoglutathione reductase (GSNOR) in plant development and under biotic/abiotic stress. Plant Signal Behav 6:789–793
12. Xu S, Guerra D, Lee U, Vierling E (2013) S-nitrosoglutathione reductases are low-copy number, cysteine-rich proteins in plants that control multiple developmental and defense responses in Arabidopsis. Front Plant Sci 4:430
13. Corpas FJ, Carreras A, Esteban FJ et al (2008) Localization of S-nitrosothiols and assay of nitric oxide synthase and S-nitrosoglutathione reductase activity in plants. Methods Enzymol 437:561–574
14. Chaki M, Valderrama R, Fernández-Ocaña AM et al (2011) Mechanical wounding induces a nitrosative stress by down-regulation of GSNO reductase and an increase in S-nitrosothiols in sunflower (*Helianthus annuus*) seedlings. J Exp Bot 62:1803–1813
15. Sakamoto A, Ueda M, Morikawa H (2002) Arabidopsis glutathione-dependent formaldehyde dehydrogenase is an S-nitrosoglutathione reductase. FEBS Lett 515:20–24
16. Moore KP, Mani AR (2002) Measurement of protein nitration and S-nitrosothiol formation in biology and medicine. In: Lester P, Cadenas E (eds) Methods in enzymology. Academic, New York, pp 256–268
17. Uotila L, Koivusalo M (1974) Purification and properties of S-formylglutathione hydrolase from human liver. J Biol Chem 249:7664–7672
18. Sanghani PC, Stone CL, Ray BD et al (2000) Kinetic mechanism of human glutathione-dependent formaldehyde dehydrogenase. Biochemistry 39:10720–10729

Chapter 16

Detection of Peroxynitrite in Plants Exposed to Bacterial Infection

Diana Bellin, Massimo Delledonne, and Elodie Vandelle

Abstract

Peroxynitrite is a highly reactive derivative of nitric oxide (NO) which is gaining attention in the plant biology community because it may play a role in NO signaling during biotic stress. Peroxynitrite can react with many different biomolecules, but its ability to nitrate the tyrosine residues of proteins is particularly important because this may regulate defense signaling in response to pathogens. The analysis of peroxynitrite levels in the context of its proposed defense role requires an accurate and specific detection method. Here, we describe a photometric assay using the fluorescent dye Hong Kong Green 2 as a specific and quantitative probe for peroxynitrite in *Arabidopsis thaliana* plants challenged with an avirulent strain of *Pseudomonas syringae* pv. *tomato*. This protocol includes the preparation of plant samples, the assay procedure, the measurement of peroxynitrite-specific fluorescence, and data presentation.

Key words Nitric oxide, Peroxynitrite, *Pseudomonas syringae*, Hong Kong Green 2, Signaling, Superoxide, Nitrosative stress

1 Introduction

Peroxynitrite ($ONOO^-$) is a highly reactive molecule that forms in vivo during the diffusion-controlled reaction between nitric oxide (NO) and superoxide (O_2^-), i.e., the chemical reaction between both molecules is so quick that rate is only limited by the diffusion of NO and O_2^- until they encounter each other in the right stoichiometry, with a rate constant (k) of 6.7×10^9 M^{-1} s^{-1}. Because NO and O_2^- are produced simultaneously in plants during the hypersensitive response triggered by avirulent pathogens [1, 2], peroxynitrite may also form and accumulate in response to pathogen infection and may therefore play a physiologically relevant role during this process.

At physiological pH levels, $ONOO^-$ equilibrates with ONOOH (peroxynitrous acid) ($pKa = 6.8$). Peroxynitrate is a potent oxidizing agent attacking different biomolecules (proteins, lipids, DNA) in the plant cell. Accordingly, early studies in animals focused on its

Kapuganti Jagadis Gupta (ed.), *Plant Nitric Oxide: Methods and Protocols*, Methods in Molecular Biology, vol. 1424, DOI 10.1007/978-1-4939-3600-7_16, © Springer Science+Business Media New York 2016

cytotoxicity [3]. However, peroxynitrite is also a strong nitrating agent, e.g., it can react with tyrosine residues in proteins to form nitrotyrosine. Proteins can be nitrated as a component of signal transduction [4, 5], thus supporting a potential role for peroxynitrite in the regulation of signaling in plants, particularly during defense responses [6]. Peroxynitrite breaks down rapidly (~10 ms) under physiological conditions into oxidizing intermediates derived from different reactions [3]. Therefore the production of peroxynitrite in plants cells undergoing the hypersensitive response is difficult to measure and until recently only limited research tools were available to monitor the accumulation of this molecule in vivo.

Nitrotyrosine-containing proteins were initially considered as markers of nitrosative stress and more specifically they were regarded as indicators of peroxynitrite accumulation. However, peroxynitrite is not the only trigger for tyrosine nitration. In animals, nitrotyrosine can also be produced via a mechanism based on heme peroxidase-NO_2-H_2O_2 [7]. Furthermore, three *Arabidopsis thaliana* hemoglobins with peroxidase activity have been shown to mediate nitrite-dependent tyrosine nitration [8]. This means that the potential role of peroxynitrite in NO signaling can only be investigated by measuring the levels of this molecule directly under physiological and pathological conditions known to induce the simultaneous production of both NO and superoxide.

As previously mentioned, given its high reactivity peroxynitrite breaks down rapidly under physiological conditions, so it cannot be quantified directly in processed biological samples. As an alternative, fluorescent probes can be introduced into living cells to report the presence of this highly reactive molecule. Among the various dyes already used to detect peroxynitrite in animal cells, aminophenyl fluorescein (APF) was the first to be used to investigate defense responses in plants, i.e., tobacco cells treated with the elicitor INF1 [9]. However, APF reacts also with the hydroxyl radical ($\bullet OH$) and hypochlorite (OCl^-) with high sensitivity, so the activity it detects is not specific.

More recently, a new BODIPY-type fluorescent probe named Hong Kong Green 2 (HKGreen2) has been developed for the specific detection of peroxynitrite. The detection mechanism is based on photoinduced electron transfer (PET) and the probe has been used successfully to measure peroxynitrite generated in activated murine macrophages [10]. We therefore established a photometric assay using HKGreen2 that allowed us to monitor the dynamic levels of peroxynitrite in *Arabidopsis thaliana* plants challenged with an avirulent strain of *Pseudomonas syringae* pv. *tomato*. This assay finally demonstrated that peroxynitrite levels increase during the hypersensitive response, strongly supporting its proposed regulatory role during NO signaling [11].

2 Materials

2.1 Plants and Cultivation Equipment

1. The method described herein uses *Arabidopsis thaliana* Columbia 0 (Col-0) plants, the accession that is used most widely for research purposes.
2. Plastic pots (9 cm diameter).
3. Plastic flats (ARAFLATs, Arasystem, Gent, Belgium) and trays (ARATRAYs, Arasystem, Gent, Belgium).
4. Soil.
5. Growth chamber with controlled light, humidity and temperature.

2.2 Bacterial Strains and Cultivation Equipment

1. The *A. thaliana* hypersensitive response is induced here using the avirulent strain of *Pseudomonas syringae* pv. *tomato* DC3000 carrying the avirulence gene *AvrB* (Pst *AvrB*). The protocol can be adapted for different pathogen strains depending on study requirements (*see* **Note 1**).
2. The following antibiotics should be prepared for selection. (a) (50 mg/mL) Kanamycin: dissolve powder in double distilled water (ddH$_2$O), filter-sterilize and store in 1-mL aliquots at –20 °C. (b) (50 mg/mL) Rifampicin: dissolve powder in dimethylsulfoxide (DMSO) and store in 1-mL aliquots at –20 °C.
3. The bacteria are cultivated in King's B broth, the following ingredients in ddH$_2$O and making topping up to 1 L: peptone 20 g, MgSO$_4$·7H$_2$O 1.5 g, K$_2$HPO$_4$ 1.5 g, glycerol 10 mL. Agar 15 g/L to prepare solid medium. The media should be autoclaved prior to use and antibiotics should be added when the media have cooled to below 65 °C.
4. Sterile 13-mL cell culture tubes with caps allowing gas exchange (Sarstedt, Verona, Italy).
5. 15-mL falcon tubes (Sarstedt, Verona, Italy).
6. Bench centrifuge for Falcon tubes (Eppendorf, Hamburg, Germany).
7. 1-mL syringe without needle (Soft-Ject, Henke Sass Wolf, Tuttlingen, Germany).
8. Incubator set at 28 °C with shaker.
9. Spectrophotometer set at fixed wavelength of 600 nm.

2.3 Peroxynitrite Detection

1. 10 mM stock solution in dimethylformamide (DMF): Hong Kong Green 2 (*see* **Note 2**). The solution can be aliquoted (*see* **Note 3**) and stored at –20 °C.
2. Cork borer.

3. 15-mL Falcon tubes (Sarstedt, Verona, Italy).

4. Vacuum source.

5. Sealed flask and connectors.

6. Flat-bottomed 96-well plates (Sarstedt, Verona, Italy).

7. Victor™ plate reader (Perkin Elmer, Waltham, MA, USA).

3 Methods

3.1 Cultivation of Arabidopsis thaliana Plants

It is important to grow healthy plants that will fully respond to pathogen infection. Most standard protocols will be effective and the following is given as an example.

1. Grow *Arabidopsis thaliana* plants in 9-cm pots filled with soil (*see* **Note 4**). Moisten the soil before sowing the seeds.

2. Sow approximately 50 seeds per pot, cover the pots with transparent film to maintain humidity and place pots in the dark for 1 day at 4 °C (*see* **Note 5**). Then transfer to the growth chamber for 10–12 days under the following conditions: 60 % relative humidity, 10-h photoperiod, light intensity 100 μmol m^{-2}s^{-1} and day/night temperature 24/22 °C (*see* **Note 6**).

3. After 10–12 days, transfer plantlets individually into trays (*see* **Note 7**) and cultivate for 6–7 weeks. Cover the flats with a close-fitting clear plastic dome to maintain humidity for the first few days and then displace the dome slightly to allow air circulation and gradually reduce the humidity. After a few days of acclimation, remove the domes completely (*see* **Note 8**).

4. Water the pots and the flats by irrigating the base.

3.2 Preparation of the Bacterial Suspension

1. One day (16–20 h) before the assay, inoculate a single colony of Pst *AvrB* from a fresh King's B plate into 3 mL of King's B liquid medium supplemented with 50 μg/mL rifampicin and 50 μg/mL kanamycin (*see* **Note 9**). Incubate overnight (at least 16–20 h) at 28 °C with agitation at 200 rpm.

2. The following day transfer the liquid culture to a 15-mL tube.

3. Pellet the bacteria by centrifuging for 4000×g at room temperature for 5 min.

4. Pour off the supernatant.

5. Resuspend the pelleted cells in an equal volume of autoclaved ddH$_2$O (*see* **Note 10**).

6. Repeat **steps 3–5**.

7. Measure the OD at 600 nm using a spectrophotometer (*see* **Note 11**) and dilute further with autoclaved ddH$_2$O if necessary to achieve a final OD$_{600}$ = 0.1, corresponding to 10^8 colony forming units (cfu)/mL.

3.3 Infection of Plants

1. Infiltrate the diluted bacterial suspension through the abaxial surface of the leaves using a 1-mL syringe without a needle (*see* **Note 12**). A mock infiltration should be carried out with ddH$_2$O as a control.

2. Place infiltrated plants under constant illumination during the infection (*see* **Note 13**).

3.4 Sample Preparation and Assay Procedure

1. At different time points after infection (*see* **Note 14**), punch eight leaf disks (5 mm) from the infected and uninfected control leaves (*see* **Note 15**) with a cork borer.

2. Transfer the disks to a 15-mL tube containing 1 mL 20 μM HKGreen2 diluted with water and infiltrate the dye into the leaf disks under vacuum for 3 min (*see* **Note 16**). Prepare a second tube containing water only and use this as an autofluorescence control (*see* **Note 17**).

3. Incubate the leaf disks for 1 h at room temperature (*see* **Notes 18** and **19**).

4. When the incubation is complete, wash the leaf disks carefully with water to remove dye solution before measuring the fluorescence.

3.5 Photometric Measurement

1. Place leaf disks individually in the wells of a flat-bottomed 96-well plate containing 100 μL of water (*see* **Note 20**).

2. Read the fluorescence emission from leaf disks every 10 min for 1 h with a fluorimeter using excitation/emission wavelengths 485/530 nm. The plate should be shaken before each measurement.

3. Calculate the peroxynitrite-related fluorescence by subtracting the autofluorescence value (from disks that were not exposed to the dye, *see* Subheading 3.4) from the fluorescence emission values of the disks infiltrated with HKGreen2 (*Pst* treated or mock infiltration) at each time point. Some representative data from one of our assays is shown in Fig. 1 (*see* **Note 21**).

4 Notes

1. Pst *AvrB* is suitable for studying the hypersensitive response in *Arabidopsis thaliana* Col-0 which expresses the RPM1 resistance protein. Because the hypersensitive response is induced following a specific race/cultivar interaction depending on gene-for-gene recognition of the avirulent pathogen, it is necessary to define suitable pathogens that can induce the hypersensitive response if other plant species are used.

2. HKGreen2 was kindly provided by Prof. Dan Yang under the terms of a collaborative agreement. More recent generations of

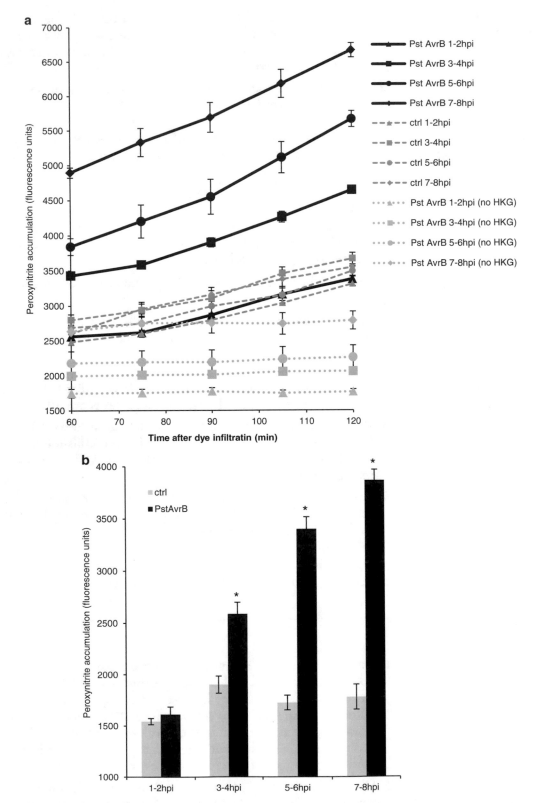

Fig. 1 Peroxynitrite formation in *A. thaliana* during the hypersensitive response induced by Pst *AvrB*. Peroxynitrite levels were estimated with HKGreen2 (20 μM) by monitoring the fluorescence intensity. Peroxynitrite

HKGreen dyes are now available such as HKGreen4, a small-molecule fluorescent probe developed for the exceptionally sensitive and selective detection of peroxynitrite in aqueous solutions, living cells and tissues [12]. These more recent dyes should be broadly compatible with the assay procedure we have described, but their chemical properties may differ slightly from HKGreen-2 and assay optimization may be necessary to find the best working solution concentrations and incubation times.

3. The dye should be divided into small aliquots to avoid freeze/thaw cycles.

4. Vermiculite and/or perlite can be added to the soil to improve aeration and inhibit fungal growth. The risk of fungal growth can also be reduced by autoclaving the soil before sowing seeds or transferring plants.

5. This step is necessary because *A. thaliana* seeds preferentially germinate after experiencing a period of cold temperatures.

6. The optimal day length and humidity depends on the species. When studying plant defense responses it is important to identify growing conditions that avoid early flowering. We recommend that *A. thaliana* plants are grown under short-day conditions.

7. Planting trays should be scrubbed to remove algae, fungi, old seeds and dead plants. Make sure the trays have drainage holes. Label the trays with the date the seeds were sown.

8. Particular care should be taken to avoid insect infestation because this may induce unanticipated defense responses that mask the response caused by the infiltrated pathogen. Pesticides cannot be used because many phytochemical products also contain molecules that induce defense responses. Therefore, infiltration should be carried out under clean-room conditions and older plants should be moved to a separate growth room for seed maturation and drying down.

9. The antibiotics used for selection depend on the choice of pathogen. *Pseudomonas syringae* pv. *tomato* DC3000 is inherently resistant to rifampicin and the *AvrB* gene confers kanamycin

---

Fig. 1 (continued) accumulation was (**a**) monitored in real time for 1 h or (**b**) estimated at different time intervals. Leaves were infected with an avirulent strain of *Pseudomonas syringae* pv. *tomato* carrying the *AvrB* gene ($OD_{600} = 0.1$) or with water (control). At different time points, leaf discs were vacuum infiltrated with 20 μM HKGreen2 and incubated with the dye for 1 h in darkness. Plant autofluorescence was estimated by measuring the fluorescence intensity of infected leaf discs that were not treated with HKGreen2 (no HKG). Fluorescence emission was monitored in a plate reader photometer for 1 h (2 h total including incubation with the dye) at room temperature. For peroxynitrite accumulation at different time points, the fluorescence values obtained at the end of the reading period (1 h) were subtracted from the autofluorescence background obtained with samples that had not been incubated with HKGreen2. In each panel, values shown are means of between six and eight biological replicates ± SE, where *asterisk* represents $p < 0.05$ vs. control. Abbreviations: *ctrl* control (H_2O), *hpi* hour postinfection

resistance. If other pathogens are used in this assay, the antibiotics should be chosen accordingly.

10. The dilution and pathogen infiltration steps do not require sterile conditions. Alternatively to water, the bacteria can be washed and resuspended in 10 mM $MgCl_2$. In this case, $MgCl_2$ is prepared as a 1 M stock solution which is autoclaved to avoid contamination. The 10 mM working solution is thus prepared freshly by diluting stock solution with autoclaved ddH_2O.

11. After growth for 16–20 h, the bacteria should reach an OD_{600} of 1.5–2. To keep within the linear absorbance range of the spectrophotometer, the bacterial suspension should be diluted 1:10 for OD measurement after washing. This should be included as a dilution factor in further calculations.

12. It is not necessary to punch holes in A. thaliana leaves to achieve bacterial infiltration. This step can be accomplished by exerting a weak pressure with a needless syringe while bracing the other side of the leaf against a finger. Special care should be taken to avoid causing tissue damage. Infiltration is complete when the leaf changes color to dark green. At least 20 µL of bacterial suspension is usually required for the complete infiltration of each leaf. Infiltrated leaves should be marked for identification because the color difference disappears soon after infiltration.

13. It is important to place infected plants under constant light because an appropriate light environment is required to establish a complete set of resistance responses in many plant–pathogen interactions.

14. Peroxynitrite is formed when NO and O_2^- come into contact. The selection of appropriate analytical time points is based on the kinetics for these two molecules, which in turn depends on the nature of the plant–pathogen interaction. In the case of incompatible interactions, it is necessary to consider how long it takes for the pathogen to introduce effectors that are recognized by plant receptors that trigger the hypersensitive response. This depends on the concentration of the bacterial suspension, with weaker suspensions inducing a slower and weaker response. A strong defense response can be ensured by using a concentrated bacterial suspension (e.g., 10^7–10^8 cfu/ mL), which can be useful to study molecular mechanisms requiring a large number of infected cells for detection.

15. There may be some variation in the intensity of the fluorescent signal among plants so we recommend the analysis of several replicates for each condition and each time point. Eight disks sampled from at least three independent plants per condition provide statistically valid data, but this number can be modified depending on specific experimental conditions. For example, if there is substantial variability among replicates then the number of leaf disks per experiment should be increased.

16. Leaf disks representing one condition should be processed simultaneously in the same 15-mL tube containing the dye working solution (or water for the autofluorescence controls). Up to four tubes can be placed simultaneously in a sealed flask connected to vacuum for infiltration, which is carried out for 3 min with continuous manual shaking to remove air bubbles. The leaf disks should settle to the bottom of the tube by the end of the process if infiltration has been successful. If there are any floating disks then the vacuum infiltration must be repeated. As an alternative, the dye can be infiltrated using a 20-mL syringe. The leaf disks are placed into a syringe containing the dye solution, which is sealed with a cap or a tip. Pressure is then exerted with the syringe. The HKGreen2 working solution can be recovered, stored at 4 °C and reused several times for a few days. The solution should be discarded if the fluorescence becomes weak or the signal-to-noise ratio declines.

17. Leaves produce many autofluorescent compounds. Notably, secondary metabolites such as polyphenols are produced during plant–pathogen interactions and these emit strong fluorescence. Because the quantity of these compounds increases during infection, a control without the dye should be used to determine background autofluorescence and should be read at the same time points to allow for background correction. The fluorescence readings should be subtracted from the values obtained from the dye-infiltrated samples in order to determine the level of fluorescence specifically generated by HKGreen2, which represents the quantity of peroxynitrite. Similarly, certain plant species show high basal levels of autofluorescence that could interfere with the experiment so appropriate controls should be set up to ensure the assay values are corrected.

18. The incubation of samples in the dark before analysis is important. Indeed, a time course of different incubation periods with plant cell extracts in the presence of the dye showed that fluorescence intensity varies with time and that incubation for 1 h is necessary for an optimal fluorescence signal. We recommend that leaf disks should be incubated with HKGreen2 for at least 1 h and the same incubation period should be used for all experimental conditions to ensure the results are comparable.

19. The stability of the fluorescent form of the dye in different plant species should be determined to ensure that the compound is not recycled by cells and that the levels of peroxynitrite measured with the dye are representative of the actual amount in the cell. For this purpose, HKGreen2 can be added to plant extracts (or water as a control) in the presence or absence of exogenous peroxynitrite. HKGreen2 fluorescence caused by the presence of peroxynitrite can then be determined by comparing the absolute fluorescence values of the plant extracts and water.

20. Leaf disks should not be damaged when they are transferred to the 96-well plates, e.g., using a flat clamp or brush.

21. The sampling time point is distinct from the start of the fluorescence measurement. This time span should be exactly the same under all experimental conditions. It is also necessary to take into account the fluorescence reading time. Data can be presented showing all recorded fluorescence values during the reading period of 1 h (Fig. 1a) or as histograms obtained by subtracting autofluorescence background obtained from disks not exposed to the dye from the signal recorded at the end of the monitoring period (Fig. 1b). In the first case, the fluorescence reading represents peroxynitrite levels in real-time for 1 h with measurements beginning 1 h after the sampling/dye infiltration time. In the second case, each bar of the histogram represents the corrected amount of peroxynitrite produced in 2 h starting from dye infiltration.

References

1. Delledonne M, Xia RA, Dixon RA, Lamb C (1998) Nitric oxide functions as a signal in plant disease resistance. Nature 394:585–588

2. Delledonne M, Zeier J, Marocco A, Lamb C (2001) Signal interactions between nitric oxide and reactive oxygen intermediates in the plant hypersensitive disease resistance response. Proc Natl Acad Sci U S A 98:13454–13459

3. Szabó C, Ischiropoulos H, Radi R (2007) Peroxynitrite: biochemistry, pathophysiology and development of therapeutics. Nat Rev Drug Discov 6:662–680

4. Gow AJ, Duran D, Malcolmc S, Ischiropoulosa H (1996) Effects of peroxynitrite-induced protein modifications on tyrosine phosphorylation and degradation. FEBS Lett 385: 63–66

5. Liaudet L, Vassalli G, Pacher P (2009) Role of peroxynitrite in the redox regulation of cell signal transduction pathways. Front Biosci 14:4809–4814

6. Romero-Puertas MC, Laxa M, Mattè A, Zaninotto F, Finkemeier I, Jones AM, Perazzolli M, Vandelle E, Dietz KJ, Delledonne M (2007) S-nitrosylation of peroxiredoxin II E promotes peroxynitrite-mediated tyrosine nitration. Plant Cell 19:4120–4130

7. Oury TD, Tatro L, Ghio AJ, Piantadosi CA (1995) Nitration of tyrosine by hydrogen peroxide and nitrite. Free Radic Res 23:537–547

8. Sakamoto A, Sakurao SH, Fukunaga K, Matsubara T, Ueda-Hashimoto M, Tsukamoto S, Takahashi M, Morikawa H (2004) Three distinct Arabidopsis hemoglobins exhibit peroxidase-like activity and differentially mediate nitrite-dependent protein nitration. FEBS Lett 572:27–32

9. Saito S, Yamamoto-Katou A, Yoshioka H, Doke N, Kawakita K (2006) Peroxynitrite generation and tyrosine nitration in defense responses in tobacco BY-2 cells. Plant Cell Physiol 47:689–697

10. Sun ZN, Wang HL, Liu FQ, Chen Y, Tam PK, Yang D (2009) BODIPY-based fluorescent probe for peroxynitrite detection and imaging in living cells. Org Lett 11:1887–1890

11. Gaupels F, Spiazzi-Vandelle E, Yang D, Delledonne M (2011) Detection of peroxynitrite accumulation in Arabidopsis thaliana during the hypersensitive defense response. Nitric Oxide 25:222–228

12. Peng T, Wong N-K, Chen X, Chan Y-K, Hoi-Hang Ho D, Sun Z, Hu JJ, Shen J, El-Nezami H, Yang D (2014) Molecular imaging of peroxynitrite with HKGreen-4 in live cells and tissues. J Am Chem Soc 136:11728–11734

INDEX

Kapuganti Jagadis Gupta (ed.), *Plant Nitric Oxide: Methods and Protocols*, Methods in Molecular Biology, vol. 1424,
DOI 10.1007/978-1-4939-3600-7, © Springer Science+Business Media New York 2016

Printed in the United States
By Bookmasters